Analytical Techniques for Food Biopreservation

Analytical Techniques for Food Biopreservation

Abeer Iqbal

Delve
PUBLISHING

www.delvepublishing.com

Analytical Techniques for Food Biopreservation

Abeer Iqbal

Delve Publishing
224 Shoreacres Road
Burlington, ON L7L 2H2
Canada
www.delvepublishing.com
Email: orders@arclereducation.com

ISBN: 978-1-77469-023-9 (Hardcover)

ABOUT THE AUTHOR

Abeer Iqbal majors in Microbiology and Molecular Genetics from University of The Punjab, Pakistan. She is a professional health-care writer and have experience of working with various life-sciences blogs. She has an international publication of her research related to microbiology. She is passionate about writing and research related to Biological Sciences. She feels that Genetics and Biochemistry are the real essence of life and learning and unraveling new researches in these areas is really amazing!

TABLE OF CONTENTS

LIST OF FIGURES

LIST OF ABBREVIATIONS

BLIS	Bacteriocin Like Inhibitory Substances
ClO_2	Chlorine Dioxide
DSC	Differential Scanning Calorimetry
ESEM	Environmental Scanning Electron Microscopy
FEO	Fennel Essential Oil
FTIR	Fourier-Transform-Infrared
GMP	Good Manufacturing Practices
GRAS	Generally Regarded as Safe
LAB	Lactic Acid Bacteria
MAP	Modified Atmosphere Packaging
MgO	Magnesium Oxide
PEF	Pulsed Electric Field
PEO	Peppermint Essential Oil
PVA	Polyvinyl Alcohol
SLNs	Solid Lipid Nanoparticles
SSOPs	Sanitation Standard Operating Procedures
TBHQ	Tertiary Butyl Hydroxy Quinine

PREFACE

Microbes frequently survive in complicated ecosystems that need them to interact with both biotic and abiotic elements. To thrive, bacterial populations must compete for space and nutrition. Microbes have developed different mechanisms, including nutrition and space competition, metabolic specialization, and cell differentiation, among others, to survive in complex and competitive environment. As a result of their normal metabolic activity, bacteria can produce a range of antibiotic compounds. Moreover, they may also develop more targeted antibacterial weaponry encoded by certain genetic determinants that are designed to destroy other microorganisms. Food spoilage is a serious problem in the food industry that causes food wastage, and significant financial losses for both manufacturers and consumers. Fungal contamination is one of the most common causes, and it can occur at any point in the food chain. Fungal growth causes food deterioration, and this will result in loss of significant odor, flavor, or texture abnormalities, but some molds also produce mycotoxin, which can have serious health consequences. Various treatments, such as fungicides and chemical preservatives, are employed to prevent microbial deterioration and hencc improve product shelf life.

Food processing is no longer as simple as it once was. It is presently transitioning from an art form to a multidisciplinary science. Several innovative preservation techniques are being developed to meet contemporary needs for cost-effective preservation and consumer satisfaction in terms of nutritional and sensory elements, convenience, lack of preservatives, low energy consumption, and environmental safety. By better controlling procedures and selecting ingredients, a better understanding and management of these traditional and complex preservation methods could aid in the development of high-quality, safe products. Food processing needs to use preservation techniques ranging from simple to sophisticated; thus, any food processing industry must acquire requisite knowledge about the methods, the technology, and the science behind the mode of action of these techniques. Keeping this in mind, this book has been developed to discuss the fundamental and practical aspects of most of the food preservation methods important to practicing industrial and academic food scientists, technologists, and engineers. Innovative technology in preservation is being developed in the food industry that can extend shelf life; minimize risk; is environment friendly; or can improve functional, sensory, and nutritional properties.

This book can be used as a guide and a handbook for students interested in pursuing a career in food science and engineering.

INTRODUCTION

In most food industries, bio-preservation and irradiation procedures are widely used. As found in most situations, traditional food preservation practices degrade nutrient content. A biopreservation technique is one of the food preservations approaches that employs antibacterial characteristics acquired from natural organisms and their metabolites. Organic acids, carbon dioxide, diacetyl, hydrogen peroxide, reuterin, and reutericyclin are some of the main antimicrobial compounds produced by bacteria throughout their metabolism. Organic acids are by-products of the fermentation process. These are organic acids that can lower the pH of the environment in the case of food biopreservation. This will also provide a selective buffer that will prevent nonacidophiles from growing. Lactic acid can damage the cytoplasmic membrane, causing other species' membrane potential to be disrupted.

Natural foods are always in high demand because they have been shown to be helpful in maintaining human health for decades. In most food processing businesses, natural preservatives have replaced synthetic preservatives. During handling and storage, bacterial growth is mostly influenced by the natural ambient conditions. Even though good manufacturing methods are practiced, food contamination by bacteria remains an issue. Food contamination has cost the food processing industry a lot of money, and unapproved food products are being sold on the market without the consent of other authorities. This approach has resulted in a slew of issues for the local economy and inhabitants. At many levels, new strategies were developed to protect the consumer's interests. Lactic acid bacteria are employed for biopreservation for a variety of reasons, including their ability to create antimicrobial chemicals that are safe for humans.

The postharvest phase can be defined as the time when food is removed from plant organs and processed in the early phases of industrial processing before being prepared for human consumption. Vegetables and fruits are living tissues that can be harvested with increased moisture content and water activity at any stage of plant development. Water is used by the collected food items, which can produce both water and heat using lipids, proteins, carbs, and other food reserves. When they are linked to plant organs, they are provided by plants in the form of nutrients and photosynthates. Most vegetables and fruits can be eaten raw, without any processing. Some foods, such as fermented items, dried foods, frozen foods, and canned meals, can be consumed in various stages and forms. The quality of raw materials might have an impact on the quality of processed vegetables and fruits.

Biopreservation has become an essential tool among food industries as most of the consumers are demanding the naturally processed foods.

Chapter

1

Introduction to Biopreservation

CONTENTS

1.1. INTRODUCTION

The foods that are derived from animals can be classified as perishable products. This can be attributed to their high nutrient content, neutral pH and moisture. It is essential to preserve food at appropriate temperatures to maintain quality and ensure safety. Bio-preservation technique can be used to increase the shelf life of several foods. This process can be achieved by antimicrobials, microbiota that are controlled and the natural biota. We can select fermentation products that are being made by beneficial bacteria (Vaishali et al., 2019). In the biopreservation (Figure 1.1) process, the pathogens will be activated, and spoilage will be reduced. Lactic acid bacteria can produce several metabolites that can be useful for the biopreservation process. The lactic acid bacteria derived metabolites present several antimicrobial properties. They can also give texture and flavor to the food products. Some of the major products or compounds produced by lactic acid bacteria are hydrogen peroxide, organic acids, and bacteriocins. Bacteriocins are nothing but proteins and peptides that exhibit antimicrobial activity.

Figure 1.1. Biopreservation.

Source: https://blog.biomall.in/bacteriocin-as-a-biopreservative-for-food-products/.

1.2. BACTERIOCINS

The bacteriocins (Figure 1.2) can be classified into four important classes (Pilar et al., 2010). These bacteriocins from lactic acid bacteria can be nontoxic and may not present any immunogenic reactions. They are thermo resistant too. Due to these beneficial features, bacteriocins can be used as biopreservative agents. We can use nisin, which is a bacteriocin in the food industries. This compound is approved by the food and drug administration of the United States. In the case of meat industries, dairy, vegetable products we can use nisin. Some of the bacteriophages can be useful in the preservation of food products. There are some endolysins that can be used in food processing industries (Hugas et al., 1998).

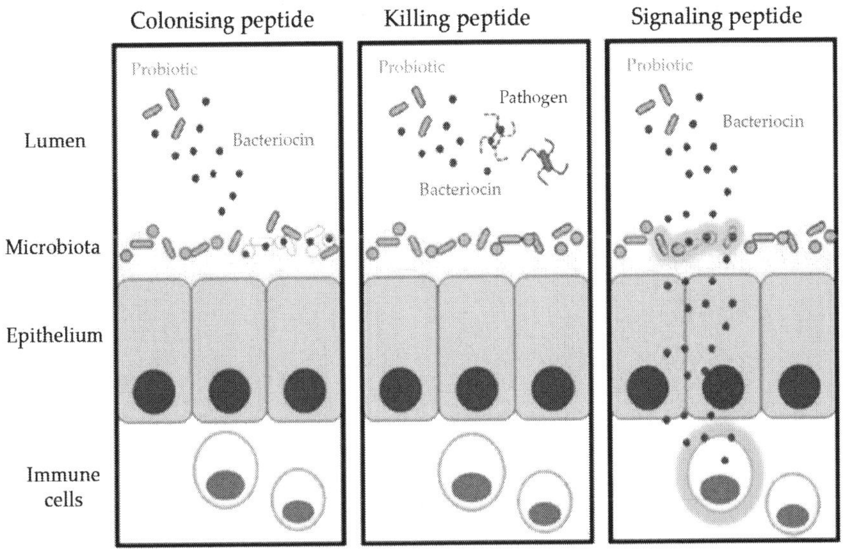

Figure 1.2. Bacteriocins.

Source: https://www.researchgate.net/publication/51753203_Bacteriocin_Produc tion_a_Probiotic _Trait/figures?lo=1.

There are disease outbreaks that can result in serious illness in humans. The illness caused due to improper management of food products has become an important and serious public issue since decades. To reduce foodborne illness, several food producing industries have upgraded their equipment so as to maintain proper hygienic practices and good manufacturing practices. The preservation of food must always be dealt with stringent measures so as to maintain quality and safety. There are several preservation techniques that mainly use low temperature like freezing and refrigeration. In some cases, we can use thermal preservation techniques such as light sterilization and pasteurization. Some of the food companies are adding chemicals to increase the shelf life of the preserved foods. There are several improvements in preservation technology pertaining to foods.

1.3. BIOPRESERVATION

Biopreservation and irradiation techniques are quite common in most of the industries. Conventional food preservation techniques reduce the nutrient quality as seen in most of the cases (Marcia et al., 2017). Biopreservation technique can be defined as one of the food preservation techniques where antimicrobial properties derived from natural organisms along with their metabolites are used. Some of the important antimicrobial substances that are produced by bacteria during their metabolism are organic acids, carbon dioxide, diacetyl, hydrogen peroxide, Reuterin and Reutericyclin. The organic acids can be classified as end products of the fermentation process. In the case of food biopreservation, these are organic acids that can decrease the pH of the environment. This will also generate a selective buffer that can stop nonacidophiles.

Lactic acid can disrupt the cytoplasmic membrane, and the membrane potential of other species will be disturbed (Michael, 1996). Carbon dioxide can be released from heterofermentative bacteria. Lactic acid bacteria can generate carbon dioxide in the process of sugar fermentation. Due to production of carbon dioxide, there will be the generation of anaerobic environment.

This will impose antagonistic effects towards aerobic bacteria, as there will be production of carbonic acid during this process. Diacetyl is another antimicrobial compound that can be called 2,3-butanedione. It can be produced as a by-product during metabolic activities of lactic acid bacteria. This compound can exhibit antibacterial activity against *Aeromonas, Yersinia, E. coli, Salmonella,* and *Listeria* species. Hydrogen peroxide is another important compound that can be produced by lactic acid bacteria when oxygen is present. It can be produced by the activity of flavoprotein oxidases (Antonia et al., 2007).

In some cases, NADH peroxidase can also help in the production of hydrogen peroxide. Hydrogen peroxide can damage proteins through the oxidation process. Due to increased membrane permeability, most of the bacterial species will be destroyed. Reuterin is a low molecular weight compound that has antimicrobial properties. It is produced by *Lactobacillus reuteri* and other lactic acid bacterial species (Marta et al., 2017). This compound can inhibit the synthesis of DNA in molds and yeasts. Reutericyclin is another important compound that can exhibit antibacterial properties. They can be produced by lactic acid bacteria. This compound can dissipate proton motive force and can regulate Gram-Positive bacterial growth. It mainly serves as H+ ionophore. Some of the Gram-Positive bacteria that can be controlled with this compound include, *Listeria innocua, Staphylococcus aureus, Enterococcus fecalis,* etc. Antifungal compounds like natamycin can be produced from *Streptomyces natalensis* species. This compound can exhibit broad-spectrum antifungal nature.

It can be used as a preservative in beverages and food industries. It increases membrane permeability of the fungal species. This will lead to leakage of peptides and ions. Due to increased leakage, there will be lysis of cells. Some of the antimicrobial compounds that can be generated from animals are lysozyme, lactoferrin, lactoperoxidase, ovotransferrin, protamine, pleurocidin and chitosan.

The lysozymes can be produced from living species as part of the defense systems. The lysozymes can be classified as GRAS (Generally Regarded As Safe) compounds, and hence they are being added to foods without any hesitation. Lactoferrin is a protein that is produced naturally from milk. Lactoferrin exhibits antimicrobial activity.

This can be due to the polycationic nature of this compound and iron-binding capacity. Because of these properties, most of the foodborne pathogens can be killed. Lactoperoxidase can be obtained from milk naturally and it kills gram negative bacteria. Ovotransferrin can be produced by hydrolysis of proteins. This compound can reduce bacterial growth by interfering with their iron absorption mechanism. Protamine can be produced from mammals, birds, fish, especially from the spermatic cells. Protamine has antimicrobial activity that kills gram positive bacteria, fungal species, and gram-negative bacteria. We can use this compound to preserve rice and fruits (Veer Singh, 2018).

1.4. YEAST

Yeast (Figure 1.3), in addition to lactic acid bacteria can exhibit biopreservation effects that can be attributed to underlying antagonistic properties. It competes with other bacterial members for nutrients. It can produce ethanol and can remain tolerant for high ethanol concentrations. It also produces a wide range of compounds that are classified as antimicrobial agents.

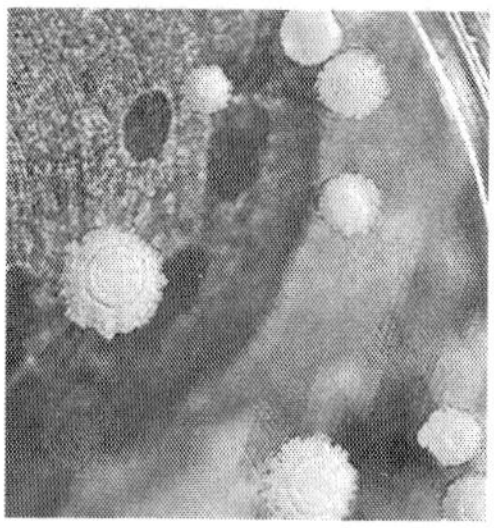

Figure 1.3. Yeast.

Source: https://www.britannica.com/science/yeast-fungus.

These compounds can exhibit a broad spectrum of antagonism in the case of bacteria. It also protects food material from human pathogens, animal and plant pathogens that may otherwise spoil the food materials. The fermentation of food by yeast alone may not be an important property to be considered it as biopreservative. In case of luxuriant growth of biopreservative bacteria, there will be gradual unavailability of nutrients for other pathogenic bacteria that might be present in the food products (Shao-Quan Liu et al., 2009). Yeast produces several antimicrobial agents that may not harm humans. Bacteriophages (Figure 1.4) are a group of viruses that can kill bacteria, and they are being used as one of the effective biopreservatives. Most bacteriophages consist of a DNA genome that is double-stranded in nature. This DNA genome is accommodated inside the capsid of virion.

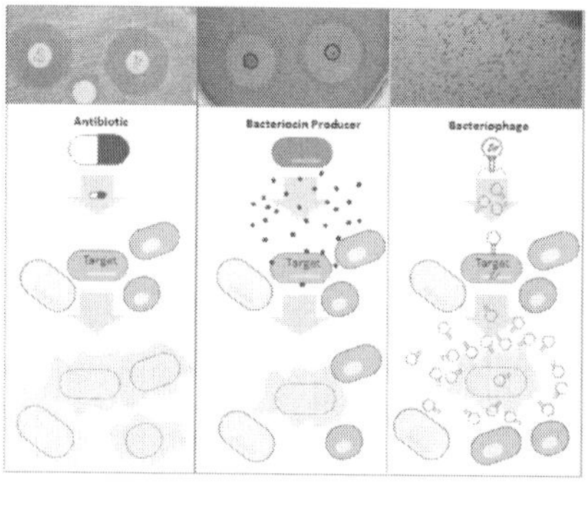

Figure 1.4. Bacteriophages.

Source: Susan Mills, R. Paul Ross, Colin Hill, Bacteriocins and bacteriophage; a narrow-minded approach to food and gut microbiology https://doi.org/10.1093/femsre/fux022.

Caudovirales is an order of bacteriophages that exhibit this type of property. These are called tailed phages. These phages can be classified into three primary families, namely Podoviridae, Myoviridae, and Siphonviridae. The phages of Podovirdae exhibit short tails, whereas Myoviridae exhibit longer tails that are straight in nature. Siphoviridae exhibits flexible and long tails. The filamentous phages contain a DNA genome that is single-stranded in nature. This DNA genome is enclosed by a protein layer. This protein layer can be helical in nature. Mostly the bacteriophages are distributed all over in nature. We can isolate bacteriophages from animals and humans to make pure cultures. In meat processing industries, biopreservation was considered as an essential aspect as meat is subjected to fermentation. The use of microbial populations extracted from animals to manage the health of animals can be called probiotics. Probiotics can be used as a tool for competitive enhancement approach. Several strategies pertaining to competitive enhancement were developed. The addition of microbial supplements are probiotics that can enhance gastro-intestinal tract microflora. Competitive exclusion can be considered as one of the essential approaches in competitive enhancement Method. These strategies mainly use the microbial system that are native to that species and organ type or system. The products generated in the fishing industry will be subjected to several contaminants due to their exposure to a variety of environments prior to reaching preservation houses. Fishing products mainly contain Omega 3 fatty acids, minerals, vitamins, and proteins along with several other essential compounds. The fish products may also intoxicate humans if not properly preserved. According to an estimate, at least 20% of foodborne problems are linked with the consumption of food products related with fish. Due to increased demands from the customers, most of the industries dealing with fishing products were forced to use conventional processes to preserve fish and derived products.

1.5. CONVENTIONAL METHODS

Some of the conventional methods that are being used by the industries are canning, smoking, and salting. Modified atmosphere packaging, vacuum packing, and low salt content-based packing were considered as mild Technologies that can be used to preserve fishing products. Cross-contamination is normal in the case of lightly preserved food products. Most of the fishing products will be consumed in their raw forms, and hence it is always essential to ensure limited microbial growth to uphold the quality and safety of the food material. The stability of food and microbial safety can be based on certain preservative factors. The texture and flavor of the fishing products are particularly important, and these properties should be taken into consideration while selecting the decontamination process. We can use ultrasound, ozone, high pressure, and pulse light in order to decontaminate fish products. The customers demand natural preservatives rather than chemical preservatives as the latter will change the native texture and flavor of the fishing products. Due to prevailing demands for fish products, the preservation techniques are upgraded to reach customer satisfaction levels. We can increase acidity in the food products by adding lactic acid, citric acid, acetic acid, etc. Some of the important preservatives that can be used are essential oils, benzoates, sorbates, sulfites, and nitrites.

As mentioned earlier, the safety of food products and their microbial stability can be dependent on different factors called hurdles. We can explain this concept with the help of the hurdle effect. This effect is being studied by most of the food industries to upgrade their equipment for food preservation. The hurdles that are present in stable products can result in required fermentation processes and in regulation of poisoning and microbial spoilage. Hurdle technology was developed and is being successfully used to reap the benefits of safe and health care food products. Homeostasis is another important aspect in food preservation technology. Homeostasis of microbes is very important in order to maintain the stability and internal environment of the organisms. It is always important to maintain optimal pH so that the organisms can be metabolically stable, and this feature can be observed in higher organisms also. If the homeostasis of microorganisms is impaired by hurdles or preservative factors that are present in foods, then the microbes may not get multiplied and hence can stay in the lag phase for prolonged periods of time. Once homeostasis gets established again, the microbes will become active and start spoiling food materials. In most of the food preservation methods, the aim of disturbing the homeostasis of microbes has become mandatory to safeguard the quality and safety of materials. Most of the hurdles in the food materials can exhibit synergist action rather than individual action. The synergic effect is more effective if the handles are acting on pH, enzyme systems, DNA, and cell membranes in the internal environment of microbes. It is always important to select a process that can be referred to as multitarget preservation of foods. Some of the hurdles are preservative factors that can be considered are preservatives like sulfites and nitrites, complete microorganisms count, redox potential, acidity, water activity, MI low temperature and high temperature. The microstructures of food, packaging methods, physical processes, pressure, and oxygen tension are also considered as important hurdles that can determine the preservation efficiency. The microstructure of foods can be linked with emulsions and solid-state fermentation. Some of the novel packing methods are enhanced edible coatings and selectively permeable films. Some other physical processes that can be classified as hurdles are ultrasonication, pulsed light processing, pulsed electric field, and ohmic heating. Radiation can also be considered as a hurdle, and it can be linked with irradiation, microwaves, and ultraviolet. Hurdle Technology approach can be used for a variety of food preservation practices. Intrinsic and extrinsic factors affect the oxidation process of membrane lipids in both animals and plants. The safety, quality, and storage life of the food products can be linked with hurdles. Most of these handles can exhibit both synergistic and additive effects. Hurdles can be classified as a set of processes are substances that can lower the deteriorative process rates. The maintenance of quality and inhibiting the deteriorative process can be considered as opposite actions in the food preservation industry. Heat processing can be considered as one of the physical hurdles that can be seen in the food preservation industry. The heat processing of foods can effectively kill the microbes and associated enzymes. It is always important to protect food recontamination. This can be done by using containers and sealed packaging. There are three main types of heat processing methods, namely blanching, pasteurization and sterilization. We can use microwave-based heating for ohmic heating in addition to conventional heating. In the sterilization process, the foot will be subjected to 200°C or above in order to destroy bacterial spores. This will ensure sufficient shelf life in case of canned foods. In the case of low acid foods that have a pH value greater than 4.5, it is important to maintain an f0 value of at least 2.5. This condition can drastically reduce spores of *Clostridium botulinum*. Sodium nitrite and salt if present in canned meat, then minimal f values can be maintained. If the pH of foods is less than 4.5, then the processing temperature can be less than 100°C. The canned foods must be stored in sealed containers. In the case of pasteurization, food can be heated to 60°C. At these temperatures some of the vegetative microbes and enzymes will be deactivated leaving the bacterial.

1.6. SPORES

Spores can survive at extreme temperatures. In most cases, pasteurization can be associated with several preservation methods like chilling and others. Hermetically sealed containers can be used to protect food materials from contamination. Blanching is another process that can be used in food preservation sectors. In this process, the temperature can be maintained between 70°C to 100°C. This technique can be used to preserve vegetables and fruits prior to processing. Temperature can be maintained before freezing and drying process.

The Global Food Industry is continuously changing its shape in order to meet the Consumer's requirements. Most consumers demand food that is safe and have a prolonged shelf life. They also demand that food is natural and free from preservatives. There should be minimal processing of food materials. Food materials should satisfy the legislative and consumer requirements, and this is possible only through sophisticated approaches in the food processing sector. This led to an increased efforts in developing food preservation systems that can process foods according to local demands. In general, biopreservation is nothing but increasing food safety and extending shelf life. Generally, the biopreservation process is dependent on biochemical reactions offered by lactic acid bacteria, and it is witnessed in the fermentation process. Fermentation can be considered as one of the oldest biotechnology related techniques that is still being used in the biopreservation process. The fermentation process provides a health friendly environment to the consumers pertaining to dietary requirements. Lactic acid bacteria can easily convert hexose sugar into lactic acid easily. This process is the main factor during heterofermentative and homofermentative metabolisms. Most of the lactic acid bacteria may not be classified under a specialized taxonomy group. This can be attributed to the fact that they are closely related as far as phylogenetic studies are concerned. They have small genomes, and this results in metabolic pathways that are quite simple and can bring about successful fermentation of carbohydrates. The lactic acid bacteria can be present in almost all places, mainly at the mucosal surfaces of several animals. Most molecular studies explain the genetic adaptation of lactic acid bacteria led to beneficial habits that can be explored by humans. We can observe simplification in metabolism that can be attributed to the reduction mechanism. Lactic acid bacteria improved duplication of genes that can help in the transport of amino acids and carbohydrate metabolism. This genetic modification has rendered lactic acid bacteria to effectively use nutrients present in milk. Lactic acid bacteria produce several antimicrobial compounds. Acetic acid and lactic acid on major antimicrobial compounds that are organic acids produced by lactic acid bacteria. We can select lactic acid bacterial strains using rapid acidification property as one of the fundamental criteria. This bacterium is essential in the dairy industry and other health industries also. We can assume that food is safe at a pH less than or equal to 4.2 and this was proved by many studies in the food processing industry. The acids produced from this bacterium can interfere with the membrane potential of cells. This will disturb several metabolic pathways. The active transport mechanisms will be impaired due to these organic acids. This bacterium can produce several metabolites that are of low molecular weight. Bacteriocins can be classified as antimicrobial peptides that are heat stable. There are several bacteriocins present in our nature. This bacterium is used as a reservoir for several bacteriocins. Most of the bacteriocins produced by lactic acid bacteria are classified as GRAS. There are several classification methods for bacteriocins. These classifications are based on post-translational modifications. Class 1 can be called as modified antibiotics, and class 2 can be called as unmodified known antibiotics. The antibiotics generated can be called lantibiotics (Figure 1.5).

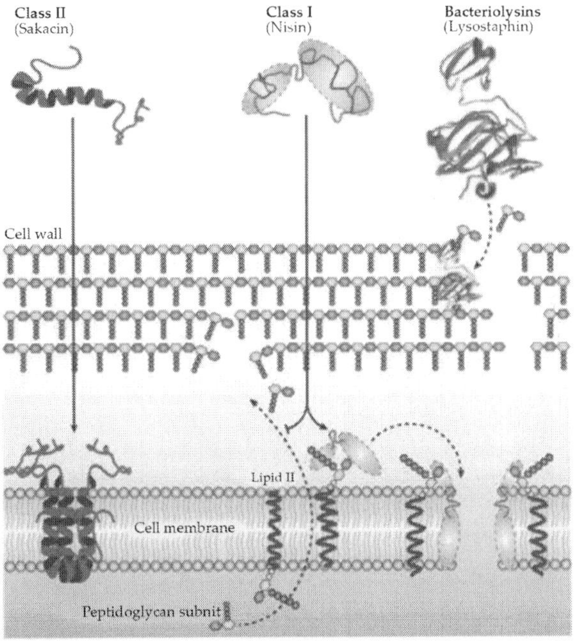

Figure 1.5. Lantibiotics.

Source: https://www.researchgate.net/publication/311946690_Isolation_of_lactic_acid_bacteria_from_milk_and_cheese_with_potential_for_food_bio-preservation_and_utilization_for_increasing_whey_digestibility/figures?lo=1.

Most of the bacteriocins classifications will be based on efflux of intracellular solutes, sugar uptake inhibition system, impairment of cell wall synthesis due to bacteriocins interference and method of membrane permeability intervention. Most of the bacteriocins may not kill all the gram-negative bacteria, and this can be attributed to the stability of lipopolysaccharides. Most of the antimicrobial activity can be linked with bacteriocin production. The bacteriocins can be used in biopreservation techniques.

Bacteriocins can withstand heat stress and they can be active at different pH conditions. They do not interfere with flavor and quality. They can be easily degraded by proteolytic enzymes, reducing the generation of resistance mechanisms. We can use bacteriocins in the bioengineering process. The activity of bacteriocins can be regulated by several factors. If the bacteriocins are unevenly distributed in the food matrix, then the activity of bacteriocins will be reduced. The solubility and slow diffusion can be considered as essential physical-chemical properties that may vary in food products, and this will affect the activity of bacteriocins. Some of the microbiota of food materials may induce antagonistic effects against lactic acid bacteria. Retention time of these compounds must be considered to know the actual activity of bacteriocins. There will be enzymatic degradation in the food products, and hence the activity of bacteriocins will be drastically affected. Functional assays were used to screen bacteriocins.

The bacteriocins can be considered as peptide and proteinaceous toxins that are produced by bacteria in order to decrease the growth of other bacterial strains. Bacteriocins share similarities with killing factors produced by Paramecium and yeast. The killing factors produced by these species can be ecologically, functionally, and structurally diverse in nature. The application of bacteriocins investigated and concluded that they are narrow-spectrum antibiotics. These bacteriocins were classified based on several metabolic activities. The bacteriocins produced by Gram-Negative bacteria can be categorized based on their size. Most of the microcin have a molecular weight less than 20 kilodaltons. We can classify bacteriocins based on genetic and biochemical characteristics. Class I bacteriocins can be called lantibiotics. They are very small and have a molecular weight less than 5 kilodaltons. They are classified as heat-stable peptides. They mainly act on membrane structures. These compounds are modified due to the translation process. There will be formation of thioether amino acids namely methyllanthionine and lanthionine. They can be produced through different steps. Most of the modifications can be witnessed after the transduction stage. Threonine and serine will suffer from enzymatic dehydration and this will result in the production of Dehydrobutyrine and dehydroalanine. Nisin can be stated as one of the examples of this process. Lantibiotics bacteriocins are classified into two main subclasses. This classification was based on structural similarities. The first class (I) comprises compounds containing positively charged peptides that are flexibly eliminated. These compounds can produce pores at the cytoplasmic membranes. Nisin belongs to subclass 1a group. The subclass Ib peptides are negatively charged, and, in some cases, there will be no charge. They are globular in structure and often maintain rigidity. They interfere with most of the enzymatic reactions that might occur in the sensitive bacterial species. Class II bacteriocin can be called nonlantibiotics. They will be having different molecular weights but not more than 10 kilodaltons. They are generally heat stable and consist of regularly occurring amino acids. This group can be primarily classified into three subgroups. They are class IIa, class IIb and class IIc. Class IIa comprises peptides that can kill *Listeria* species.

1.7. ESSENTIAL OILS

Essential oils (Figure 1.6) can be classified as natural substances that can be utilized as natural food conservators. It is always important to investigate the complete biological properties of these oils. Oil from Hyptis species can increase the shelf life of beef. This Essential oil can be extracted from plant materials with the help of hydro distillation units.

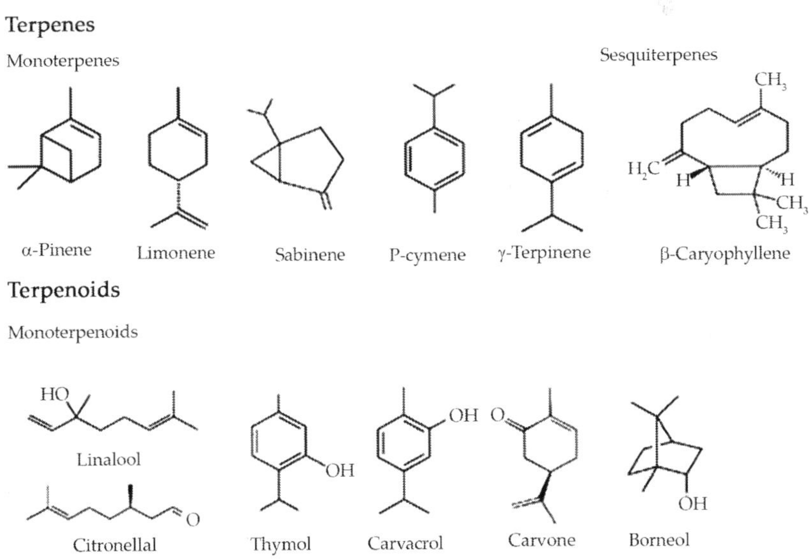

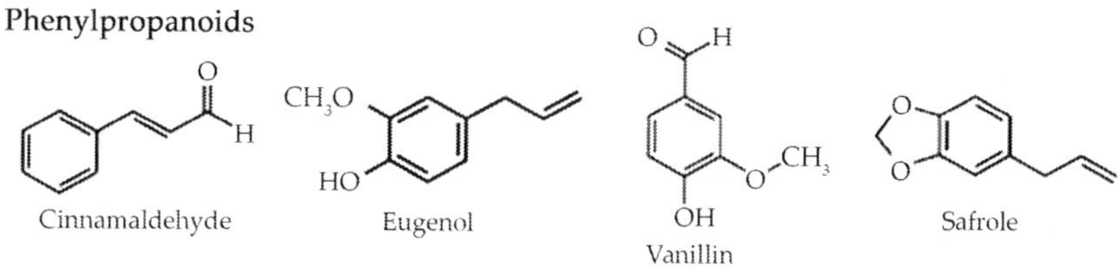

Figure 1.6. Essential oils.

Source: https://www.researchgate.net/publication/221792983_Essential_Oils_in_Food_Preservation_Mode_of_Action_Synergies_and_Interactions_with_Food_Matrix_Components/figures.

The ground beef and meat will be infected by food poisoning agents that can compromise food safety parameters. Microbiological safety of the food materials can be considered as an adynamic entity, as the safety of the food will be affected by several factors per se. Most of the pathogenic microbes affect the food safety parameters, and we can consider meat as an excellent vector that can transmit several pathogens such as parasites, prions, viruses, and bacteria. The host will be colonized with this bacterial or any pathogenic species on consumption of this meat. There are several modern technologies that are being used in food processing industries to eliminate several pathogens. Food spoilage is nothing but any damage that can be caused to devastate the actual nutritional value of the food materials. Due to food spoilage, flavor, texture, and other properties will be affected, and that can harm consumers. There is increased consumption of precooked foods all over the world, and this is mostly restricted to seafoods. The temperatures are not being properly maintained during its storage and hence can pick up certain pathogens enroute. This is most commonly seen in developing countries that lack proper infrastructure to store food materials for extended periods of time. There are some countries that cannot support sanitary requirements during transit of these food products. In the United States, at least 250 million people suffer from gastroenteritis, and that results in 500 deaths every year. Most of these problems can be attributed to food borne diseases caused by dairy products, seafoods, eggs, poultry, and meat. Some of the popular bacterial pathogens that are related to food-borne illnesses in humans include for *Clostridium botulinum, Staphylococcus aureus, Listeria monocytogenes, E. coli, Campylobacter jejuni* and *Salmonella species.* Most of the food industries rely on physical treatments and chemical preservatives in order to store food materials for extended periods of time. Preservation techniques are not new as they have been used for centuries. Some of the chemical preservatives can induce toxic effects in human beings. This can be attributed to changed nutritional properties and organoleptic properties of the food materials. It is always recommended to consume food without additives and that are minimally processed. Due to increased demand for preservatives free food and additives free food has resulted in the use of natural products in the food industries, at least for now. Biotechnology is being used in several food industries, ensuring that the microbes can protect the food materials from being spoiled. There will be improvement of useful microorganisms and derived products so that we can use these products to protect the food quality. The metabolites and the nonpathogenic microbes can encourage microbiological safety along with prolonged shelf life of these foods, and this was called as biopreservation. The lactic acid bacteria exhibit several antagonistic properties and was proved in several areas of food industries. There are certain strains that can produce bacteriocins. These strains can be used as protective cultures to safeguard the food products but without harming humans. There are certain characteristics that are to be maintained while using these organisms as food preservatives. These organisms should exhibit broader bactericidal activity, and using these trends should reduce both gram-positive and gram-negative bacteria that are otherwise pathogenic in nature. We need to use certain strains that are heat resistant as the food products will be subjected to sterilization and pasteurization. The microbial strains that are selected should not act on eukaryotic cells. They should not be toxic to the laboratory animals and should not exhibit any type of immunogenic reactions. They should get inactivated by the enzymes produced in the gastrointestinal tract of human beings. We need to investigate the genetic determinants that might be present in the plasmid, thereby permitting us to alter the genetic setup of the microbes. This will lead to the generation of natural peptide analogs with beneficial features. The strains that produce bacteriocins must be selected carefully to ensure your safer food applications.

Adding bacteriocins producing strain to the food should have no adverse effects on the food nutrients and quality, in fact, they should have improve safety along with beneficial effects. Most of the producer's strains must be previously declared as GRAS species. The bacteriocins exhibit cytolytic activities. Recombinant DNA technology is being used to improve the effectiveness of producer's strains. Bacteriocin producing genes can be transferred to other species in order to increase broader activity. It is important to study the functional and structural aspects of bacteriocins. Bacillus species can produce bacteriocins. This species is considered in the laboratory and bacteriocins will be isolated and characterized. *Bacillus subtilis* species can generate subtilin. It is classified as a cationic peptide. It has a molecular weight of 3317 daltons. It belongs to antibiotics group of bacteriocins. Bacillus species state-certain benefits along with lactic acid bacteria. Although it has several beneficial features, there are certain problems with this species as some members of this genus are creating certain problems in human beings. The bacteriocins generated from this bacillus species can be regarded as safe. Bacillus species can exhibit antimicrobial properties against Gram-Negative bacteria and Gram-Positive Bacteria. Some of the bacillus species can kill fungal members. Bacillus species can exhibit wider antimicrobial spectra to that of the lactic acid bacterial spectrum. We can observe several metabolic diversities that can yield bacteriocins with different properties. These bacteriocins can exhibit inhibitory activity at both basic and acidic conditions. Bacteriocins can exhibit inhibitory effects even after thermal processing of food products. They can be useful in food processing industries. The genetics and physiology of bacillus species is well researched and documented. There are several molecular biological tools that can be used as screening techniques in the food processing industry. Bacteriocins can be developed by both Gram-Negative bacteria and Gram-Positive Bacteria. We can observe great diversity with bacteriocins. Bacteriocins that are produced by lactic acid bacteria have got a lot of advantages over others. There are several approaches in the application of bacteriocins to preserve food products. We can use products that are fermented previously along with bacteriocin producing microbes in the food processing industry. We can add semi purified bacteriocins as food preservatives, and in some cases, we need to add purified bacteriocins. The food can be inoculated with lactic acid bacteria that can produce bacteriocins as products. The rate at which lactic acid bacteria grow and produce bacteria can be crucial in defining the food preservation process.

Biopreservation generally refers to enhanced safety of foods and extended storage life with the aid of natural microbes or their derived products. It is noteworthy to mention that lactic acid bacteria can be easily used in the biopreservation process as it is safe to consume and mostly it dominates the native microflora of the food materials that are to be preserved. In the case of meat, cereal products, vegetables, and milk, lactic acid bacteria can change the nature of food products according to the customer requirements. Fish and raw meat can be stored in freezing temperatures, preferably under vacuum. They can be stored in increased carbon dioxide concentrated areas.

In almost all cases, the lactic acid bacteria dominate other microbes that can result in hidden fermentation.

1.8. FUNGAL CONTAMINATION

The loss of food has become an important concern all over the world due to exponential population growth rates. Food quality can be damaged by fungal members like Penicillium and Aspergillus and others. The secondary metabolites produced by these fungal members can cause toxic effects on animals and humans. These are named mycotoxins (Figure 1.7).

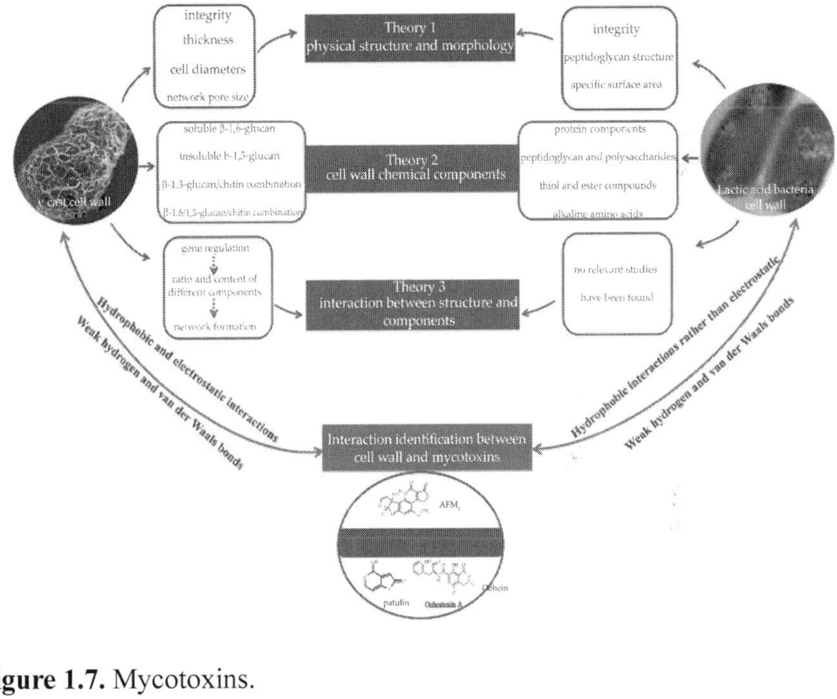

Figure 1.7. Mycotoxins.

Source: https://www.sciencedirect.com/science/article/abs/pii/S0278691517307494.

The mycotoxins are classified as heat-stable compounds that can withstand extreme temperatures during food processing steps. This is of important concern in most of the food preservation industries as it will compromise the food safety protocols. Fungal spores can easily be propagated through sexual and sexual reproduction methods. Once dispersed that can easily get settled in the food products and extend their lifespan. Fungal spores can attack food products at the field harvest stage, postharvest stage, processing stage, and storage stages. The food materials can get spoiled due to fungal spores even a wholesaler, retailer or at consumer levels. There will be severe producer losses due to fungal infestation, and this will bring about a negative impact on the food distribution companies. It is always important to regulate fungal contamination at each process of food storage. There are several adverse health effects and environmental risk situations that may arise due to the application of preservatives and fungicides. This has forced many companies to opt for natural methods of preservation without changing the nutritional characteristics of food. The fungal members have developed a resistance against several preservatives and pesticides. Because of the nutrient rich ingredients in the food materials, fungal spoilage is prevalent in most of the industries. The spoilage of foods due to fungal infection can be regulated by several factors. Before using biocontrol tools and biopreservation tools, it is always important to know the fungal contaminants that dominate the microflora of the food products, especially the packaged foods. The identification and isolation of the fungal species is always important so as to ensure your best management practices in the food processing industry and food preservation sector. The mycotoxigenic potential should be defined in the selection process to select appropriate microbial candidates to reduce food spoilage. The identification of the fungal species must be thoroughly done using appropriate methods. We can identify different molds using Pitt test, and through this test, we can easily know the morphology of the fungal members. Several molecular methods were developed to identify fungal species accurately. PCR amplification technique have been developed, and it is being used to know the properties of ribosomal subunits in the fungal members. The beta-tubulin gene is crucial in order to identify some species. Vegetables, fruits, and other raw materials can be susceptible to fungal contamination in their respective ripening stages. This can be attributed to changes that might occur in pH values, production of carbohydrates, morphological changes in skin and disrupted defense mechanisms are weakened defense mechanisms. The fungal species can induce several symptoms in postharvest crops. There will be the formation of tissue lesions and discoloration of tissues in the crops infected with fungal species. The fungal members can spoil a wide variety of vegetables and fruits. Some vegetables such as legumes, crucifix and tubers are least affected by fungal pathogens than fruit bearing plants. The pH conditions in the fruit bearing plants encourage the growth of fungal pathogens such as mucor, gloeosporium, fusarium and aspergillus species. We can observe Alternaria, Rhizopus and Penicillium species in the food products. The diseases caused by this fungal species can be called postharvest diseases. Fish, fresh milk and the raw materials produced from animals are least susceptible to this point is caused by fungal members, but they are sensitive to bacterial spoilage. Most of the food spoilage due to fungal members may occur during refrigeration, and this does not change the bacterial population that is already predominant in the food products that are being refrigerated. Most heat related food processing treatments can permit fungal destruction but, in some cases, this is not clear. Some of the ascospores that are being generated by fungal species are resistant to heat treatments, and this leads to spoilage of fruit juices and few fruit purees.

The biocontrol or biopreservation can be used to regulate the native microbiota that may spoil the food products. Most of the products generated by the beneficial microbes can be antibacterial in nature. Fungal diversity has helped many food processing in exploring different food products. There are two essential food products specified by Pitt and Hocking. They are perishable foods and fresh foods and processed food, and stored foods. The fungal species are being cultured using agar plates. Latest high throughput assays were developed using multi-well plate technique. These techniques were used to investigate the microbes via screening process. Some of the pure fungal cultures were isolated using these techniques. These pure cultures can be used to inhibit the fungi that spoil the food products. The supernatants and cultures of the beneficial strains can be grown on the agar plates according to the requirements. The inhibition halo produced in the agar plates can represent the antifungal activity of the fungal strains. There are certain specialized techniques called dual culture techniques. This can be used to bring about antifungal activity or antagonistic activity. In this technique, the antifungal molds can be inoculated over both sides of the agar plates. Glucanases, chitinases, proteases and cellulases can be produced from the fungal species and they can be used as cell wall degrading enzymes. The agar plate methods were upgraded with 24 wells in order to enhance the screening process.

There are several in vitro screening experiments on methods that can be used to investigate the antifungal properties of lactic acid bacteria. Most of these experiments were prepared using synthetic media like MRS Agar medium, Rogosa and de Man medium. The ingredients of MRS Agar medium can affect the antifungal activity imposed by lactic acid bacteria. In this medium there will be acetate and other important ingredients that can help in strengthening the antifungal activity. Some of the antifungal compounds that can be produced by these fungal members and lactic acid bacteria can be classified as safe and can be used in several food processing industries. Raw meat products can be loaded with microbial populations, and these populations are dominated by Enterobacteriaceae members and lactic acid bacteria. The metabolism of microbes can bring about spoilage to all the meat products if not properly maintained. It harms humans through several ways. The pathogenic bacteria that is normally present in the meat products can be increased in numbers during the refrigeration stage. Listeria monocytogenes is one such bacterial species that can spoil meat products on refrigeration. The toxin producing bacteria should be regulated continuously. We can decontaminate the surface of raw meat using nisin combinations with MAP, vacuum packaging, lysozyme, chelators, and organic acids. We can reduce lactic acid bacteria and Carnobacterium by using nisin activated films. We can use pediocins to protect raw meat products. Pentosin can be used to reduce the growth of Pseudomonas and listeria species. In some food industries, Pulsed light, Along with nisin is used to control several bacterial species in the meat products. A lot of research is being performed that does not know the exact nature of bacteriocins in preserving raw meat, and most of these products can be safeguarded by nisin. There are certain drawbacks with this compound. It has limited solubility rates, and it gets interacted with several phospholipids. It can also exhibit antagonistic properties with glutathione. This compound can successfully decontaminate the surface of RAW meat. This compound can be mixed with polylactic acid and lactic acid in the solutions and the meat that is to be sterilized or decontaminated Can be brought into contact with this solution. We can use nisin with EDTA with vacuum packaging to clean the raw poultry meat products. We can use other bacteriocins like pentocins, bifidocins and carnobacteriocins to inhibit several pathogenic bacteria. Another important approach to regulate the activity of bacteriocins in fresh meat products is their immobilization in the substrates. The mobilization in substrates mainly used coatings, beads, and liposomes. Most of the meat products that are cooked can be classified as ready to eat foods and that consist of meat pieces that might have been generated by the grinding process. In the cooking process, the natural biodata will be killed to some extent. There will be post-process contaminants in the food materials that are already cooked, and these are to be investigated to a deeper extent possible. We need to consider pH values of the cooked meat products as the pH values post-cooking will favor the growth of pathogenic bacteria. There are certain meats that should be subjected to further processing like packaging, peeling, and slicing. There will be cross-contamination during these processes. The lactic acid bacteria strains can be classified as major food spoiling bacteria. These bacterial species can be found in vacuum-packed meat products. These species can produce gas and Slime during their preservation. Ovotransferrin can be considered as an important antimicrobial compound present in the egg of hens. The antimicrobial activity of this compound can be attributed to its iron-binding capability. This chemical can easily bind with bacteria on its surface and inhibits their growth to a large extent. There are some lactic acid bacteria species that can be used in the fermentation of meat. It is always essential to use strains that produce bacteriocins well before the start of the inoculation process. Milk can be categorized as an important vehicle or vector that can spread pathogenic bacteria in human beings. It is essential to opt for a pasteurization process before consumption of raw milk. Due to this process, the microbiota that spoils milk will be regulated, leaving a certain proportion of bacterial species behind. Several bacteriocins protested on dairy products and Milk products in the food preservation industries. These bacteriocins were used to inactivate pathogenic bacteria or bacteria that is involved in the spoilage of milk and its dead products. The shelf life of milk can be increased by the addition of nisin compound. These compounds are essential to protect milk from getting spoiled during transit for longer distances. The heat treatment along with the application of nisin compounds can effectively decrease the D values of bacillus species and other natural microbiota that can spoil milk. Lacticin can be used to preserve diary food products for extended periods. These compounds primarily act on cytoplasmic membranes of bacterial cells and alter permeability rates effectively. Due to altered permeability rates, most of the bacterial species will get killed. These compounds can initiate cell lysis. The bacteriocins can also preserve eggs and derived products. We can extend the shelf life of eggs and regulate the pathogens that are present in it by the addition of these compounds. Aseptic condition should be maintained during packaging to prevent contamination after heat treatment.

Chapter

2

Antimicrobials

CONTENTS

2.1. INTRODUCTION

There is always a growing demand for natural foods as they are proved to be effective to sustain human health for decades. Natural preservatives have replaced the synthetic preservatives in most of the food processing industries. The bacterial growth is primarily dependent on the native environmental conditions while handling and storage. Good manufacturing practices were allowed, still there is a problem of food contamination by microbes. The food contamination has resulted in heavy losses in food processing industries, and unauthorized food products are being supplied in the market without approval of other respective authorities. This practice has led to several problems with the local economy and population (Nychas, 1995). New techniques were developed to safeguard the interest of the consumer at various levels. Lactic acid bacteria are being used for several purposes of biopreservation due to its capacity to produce antimicrobial compounds that do not harm human beings. Bioactive compounds can be produced by *probiotic bacteria*, and some of them are exopolysaccharides, immunomodulators, alcoholic compounds, anti-inflammatory agents, antioxidants, vitamins, organic acids, fatty acids that are having short chains, peptides, amino acids, enzymes, and metabolic bacteriocins. The submerged mode of fermentation has been used for centuries to ensure sterilization. This technique is being used to maintain balance in the medium and heat transfer process. This technique can be applied in case of lactic acid bacteria that can easily grow in high moisture areas. It is to be noted that the biosynthesis of bacteriocins cannot be equal to bacterial growth per se (William Strohl, 2014).

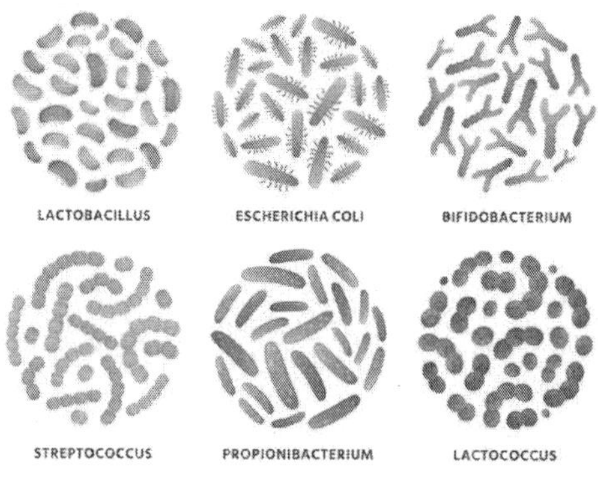

Figure 2.1. Probiotic bacteria.

Source: https://www.freepik.com/premium-vector/lactobacillus-bifidobacterium-lactococcus-probiotic-bacteria-circle-good-bacterias-illustration-set_13218919.htm.

2.2. PROBIOTICS

The *probiotic bacteria* (Figure 2.1) may produce vitamins, fatty acids, and amino acids. The production of bacteriocins can be higher if the nutrients, pH, and temperature are at minimum for probiotic strains. The production of bacteriocins has attracted several researchers as it has proven to be an amicable option in the current food processing scenarios (Erika et al., 2004). The mass production of bacteriocins is limited by increased expenditures and the limited yield of peptides. The minimization of cost can be achieved by optimizing the whole fermentation process and by utilizing bioengineering strains that can produce increased amounts of bacteriocins. We can use organic residues in the fermentation process as essential substrates, and this will reduce several monetary inputs to the production of bacteriocins. There are certain bacteriocins that can be partially purified with the help of Sab states like milk, and they were classified as Food grade substrates. Industrial and agricultural waste can be classified as nutrient-rich substrates because they contain both inorganic and organic materials (Nancy, 2010). We can use these substrates as alternative inputs of both nitrogen and carbon. These products can be easily used by several microorganisms that can be both beneficial and user-friendly. These bye products can be produced by Dairy industries at a large scale and that can be easily used in the fermentation process. Recycling and using this byproducts can reduce the environmental impacts to a greater extent. In the case of cheese production, there will be an increased generation of whey. It contains lactic acid, proteins, salt, and lactose. pH values, proteins, minerals, vitamins, lactose, and other essential compounds are present in this byproduct, and this can act as an excellent substrate for several microbes that need neutral conditions. We can also add sucrose to these substrates. Lactic acid bacterial strains can easily produce antimicrobial bacteriocins, but that production will be dependent on bacterial growth conditions, pH, temperature and the strain that is being used for their production. With sodium chloride concentration can be treated as an essential factor in culturing lactic acid bacteria strains (Afrc, 1989). The increased concentration of sodium chloride can also bring down the antimicrobial activity by other species. Sodium chloride can also disturb the binding capacity of bacteriocins on the receptors that are having delicate cell membranes. Temperature is another important factor that can increase microbial growth rates. The pH of the medium can control the Migration of several components and activity of enzymes across the cell membrane. The production of Sakacin P is high at 20°C. Although several bacteriocins are being identified, the purification of these compounds has been a big problem for most of the industries. The purification process of bacteriocins can be classified as a slow process. In this process, there will be precipitation, filtration, and centrifugation. The precipitation process will be carried out through Ammonium Sulphate. There will be solvent extraction and chromatography in the purification process. *Lactobacillus plantarum* can exhibit antibacterial activity on both Gram-Negative bacteria and Gram-Positive Bacteria. Biochromatography along with HPLC is being used to purify bacteriocins produced from this bacterial species. We can use reverse-phase high-performance liquid chromatography in this process. There is a need for purification methods that are simple and economical in nature so as to enhance viability among the industrial counterparts (Arakawa et al., 2010).

2.3. INOCULATION

There are several studies that represent the search for cheaper alternatives for high-cost media that is being used for the production of bacteriocins. It is always essential to produce cost-effective bacteriocins because of the benefits that might be reaped in the food processing sector. Most of the antimicrobial components are compounds that can either kill or delay the growth of undesirable microbes in the food products. We can classify antimicrobials into synthetic antimicrobials and natural antimicrobials. The bacteriocins that are being used in the biopreservation sector can be used in foods in several ways. We can easily incorporate bacteria onto food products with advanced techniques.

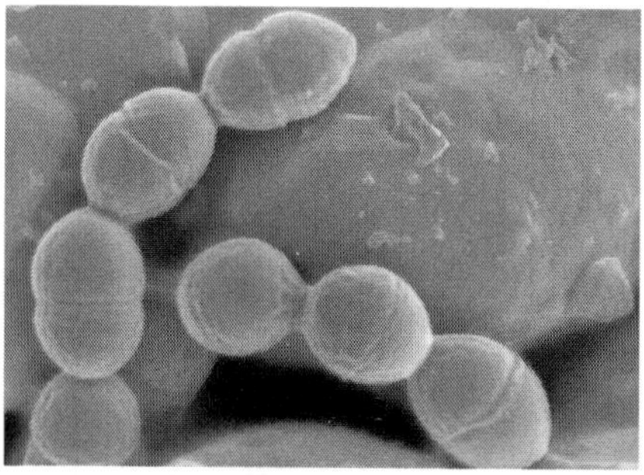

Figure 2.2. Lactococcus lactis.

Source: http://textbookofbacteriology.net/featured_microbe.html.

We can add partially purified bacteriocins into the food products. We can also add the fermented products of the bacterial strains that can produce bacteriocins. The bacteriocins produced by *Lactobacillus plantarum* can be added to raw beef to increase its shelf life (Amrita et al., 2019).

Lactococcus lactis (Figure 2.2) derived bacteriocins can be applied on jerky beef in order to reduce decomposition rates. Pediococcus pentosaceus (Figure 2.3), Staphylococcus xylosus (Figure 2.4), and Lactobacillus sakei (Figure 2.5) can produce Bacteriocins that can protect pork sausage. The bacteriocins produced from these species can preserve food products then chemical preservatives. Lactobacillus curvatus derived bacteriocins can preserve pork that is cooked, and there will be inhibition of listeria monocytogenes.

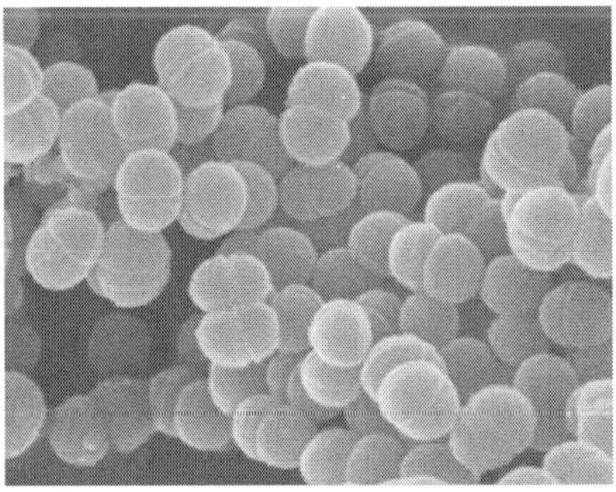

Figure 2.3. Pediococcus pentosaceus.

Source: https://sfamjournals.onlinelibrary.wiley.com/doi/pdf/10.1111/j.1365-2672.2008.03770.x.

Figure 2.4. Staphylococcus xylosus.

Source: https://pixels.com/featured/staphylococcus-xylosus-scimat.html.

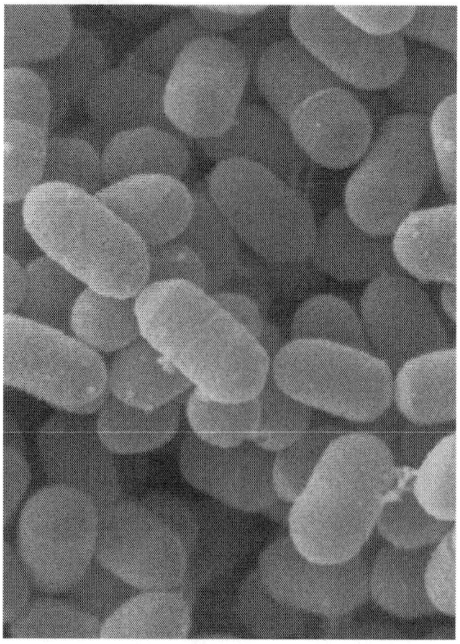

Figure 2.5. Lactobacillus sakei.

Source: https://fineartamerica.com/featured/2-lactobacillus-sakei-sem-scimat.html.

The bacteriocins produced by *lactobacillus Sakei* can preserve Salami by inhibiting *Listeria monocytogenes*. *Weissella hellenica* can produce bacteriocins that can produce fish filet by inhibiting different bacteria that may spoil fish products (Delves et al., 1996). The bacteriocins produced by *Enterococcus mundtii* can produce fish by inhibiting Gram-Positive Bacteria. *Lactobacillus curvatus* derived bacteriocins can produce fish products by inhibiting mesophilic bacteria and several coliforms along with listeria species. *Lactobacillus acidophilus* and *Lactobacillus plantarum*, along with Streptococcus thermophilus species can produce bacteriocins that can be used to preserve food products pertaining to fish by increasing the shelf life of the fish. *Lactobacillus reuteri* derived bacteriocins can be used to protect fish filet by inhibiting diverse bacterial species. *Lactobacillus plantarum* can be used to preserve chilled shrimp by inhibiting *Vibrio parahemolyticus* true bacteriocins. *Lactobacillus rhamnosus* can produce bacteriocins the temperature of goat cheese by inhibiting *Salmonella, staphylococcus, Listeria,* and *E. coli* species. The bacteriocins produced from *Lactobacillus brevis, Lactobacillus plantarum,* and *Lactobacillus casei* can be used to preserve Himalayan cheese by inhibiting diverse microbes.

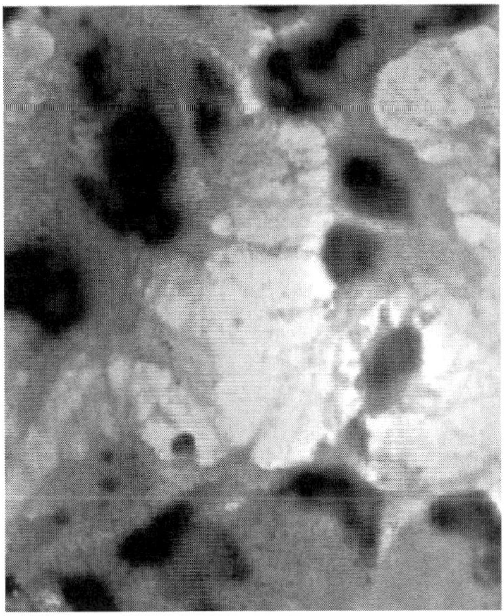

Figure 2.6. Lactobacillus reuteri.

Source: https://www.nature.com/articles/s41598-018-35702-y/figures/1

Lactobacillus reuteri (Figure 2.5) can produce bacteriocins that can effectively preserve semi hard cheese. This can be done by inhibiting *E. coli* and *Listeria monocytogenes. Pediococcus pentosaceus* can be used to produce bacteriocins that can protect vegetables by inhibiting listeria monocytogenes. *Lactobacillus lactis* can protect the lettuce by producing bacteria since that can inhibit *Listeria monocytogenes. Lactobacillus pentoses* can be used to reserve cut fruit by inhibiting *E. coli, Salmonella* and *Listeria monocytogenes* species. Reactivity of bacteriocins as biopreservation to agents can be dependent on the biosynthetic pathways that are present in microbes that live in food products. The specificity of microbial action is also important as the food matrix plays an important role in determining the specificity of antimicrobial compounds. These bacteriocins can affect the organoleptic properties of food. They may get chemically degraded and physically altered with Limited water solubility when present in food products that are to be processed. It is crucial to evaluate all the structural and functional parameters of bacteriocins before adding them to food products that are to be preserved for a certain period of time. The bacteriocins that are being applied to the food products should not affect human health at any scale (Marta et al., 2017).

In recent years, there was a lot of research that focused on bacteriocins that might act as biopreservation agents. Several studies well done on the fermentation process of pork sausages. The experiments were done using *Staphylococcus xyloses, Lactobacillus sakei,* and *Pediococcus pentosaceus* along with naturally fermented sausages in order to understand the extent of biopreservation. The sausages that were inoculated with this microbial species exhibited minimal nitrate levels that can be called as residual nitrate levels. The spontaneously fermented sausages exhibited high residual nitrate levels. Inoculated sausages exhibited lower levels of malondialdehyde, whereas spontaneously fermented sausages exhibited increased amounts of malondialdehyde. The sausages that were inoculated represented limited oxidative stability with nitrite depletion. This property can be explored to enhance the hygienic quality of the fermented sausages, thereby increasing the food safety of them. In the case of sausages, the flavor that will be altered because of inoculations can be a desired factor as far as the customers are concerned. According to a study, jerky beef exhibited negative effects on microorganisms that are tolerant to high salt concentrations. The bacteriocins are produced by *Pediococcus* species. We previously tested with the help of different division tests. This bacteriocin were compared with sodium benzoate and sodium sulfite in order to know the actual biopreservation potential of bacteriocins. 100 micrograms per liter of bacteriocin extract from this bacterial species exhibited antimicrobial properties against *Shigella* species and *E. coli*. In the maturation of cheese, the microbiota plays an essential role in the production of aromatic compounds. The microbiota can help in the conservation of lactate, lactose fermentation, lipolysis, proteolysis, and several other reactions. The cheese maturation can be regulated by different fungal species including yeast. This maturation by fungal members can carry both positive and negative impacts on the flavor of cheese. The decoration of cheese can be due to the presence of molds. These molds can produce mycotoxins and false pigmentation in the cheese. Spoiling cheese my fungal members can render it an answer table for human consumption because of bitter flavor and other changes in cheese. This had led to severe economic losses in the cheese producing industries. Fusarium species can readily attack cheese and spoil it. Some of the Other fungal species that can damage Cheese are *Aspergillus*, *Penicillium geotrichum, Mucor,* and *Cladosporium* species. The deterioration of cheese can be lowered by chemical preservatives and pasteurization. Some of the microbes can easily disintegrate via the decarboxylation process.

There are many approaches that can be used to reduce the growth of molds in the cheese products. The propionic acid bacteria and lactic acid bacteria can be used as best candidates in the case of fermented foods (Schwenninger et al., 2011).

When Lactobacillus species is added to the goat cheese, we can regulate the growth of pathogenic strains like *Listeria monocytogenes, Escherichia, Salmonella Enteritidis,* and *Staphylococcus aureus* species. We can add probiotic strains to goat cheese in order to protect it from the *Listeria* and *Staphylococcus species*. Goat cheese can carry several probiotic strains like *Bifidobacterium lactis, Lactobacillus casei,* and *Lactobacillus acidophilus* species. Listeriosis can spread due to the consumption of the raw vegetables. We can opt for several rounds of washing treatments, preferably with the antimicrobial agents to lower the microbial populations. Chlorine can be used as a best disinfectant in case of vegetable cleaning. Chlorine-based disinfectants can easily produce several toxins that can cause health problems in humans.

The demand for seed sprouts, radish, soybean, alfalfa, and mungbean is on the rise due to potential health benefits offered by them. These vegetables can be used by the customers in their raw conditions in the form of sandwiches and salads. The seed sprouts have been used since ages as a commodity in desserts. Japan is one territory where the consumption of seed sprouts is high. Due to their potential health benefits, they are being used by many countries all over the world. The seed sprouts of wheat, sunflower, sesame, rye, rice, radish, onion, yellow and green pea, white mustard, soybean, mungbean, lentil, cress, broccoli, alfalfa and adzuki bean are popular. As these seed sprouts are consumed in raw conditions, there are several health concerns. Enterobacteriaceae and Pseudomonas can be carried by these raw seed sprouts. Initially, the bacterial load will range from 3 to 6 log CFU/gram and during the sprouting process, there will be an increase in the bacterial load. We can expect 8 log/CFU per gram. Diverse bacterial strains were identified in these seed sprouts. Some of them are *Vibrio cholerae, Yersinia enterocolitica, Bacillus cereus, Staphylococcus aureus, Listeria monocytogenes, Salmonella spp*., and *Aeromonas hydrophila* species. The health problems associated with the Salmonella stereotypes are being associated with *E. coli* and *B. cereus*. Most of the outbreaks are linked with Alfalfa sprouts. It is noteworthy to understand the microbial antagonism mechanisms that can be an amalgamation of mechanisms. These mechanisms are regulated by the competition for nutrients and space. Environmental variations can also affect the microbial antagonism mechanism. The bacteriocins, lytic enzymes, antibiotics, oxygen radicals, organic acids can alter the course of microbial antagonism.

2.4. BACTERIOPHAGES

The lytic bacteriophages (Figure 2.6) can be used as biocontrol agents as they are species specific in nature and their target only pathogenic bacteria. Beneficial bacteria will not be affected by bacteriophages. Because of this unique property, bacteriophages are being used in the food industry (Jiai et al., 2021). Lytic bacteriophages are self-propagating in nature. This can be an advantage in case of inoculation within the food safety standards. Will also decrease the processing cost and hence industries can easily opt for bacteriophages. It is important to add complex phage mixtures in the food processing industry, and this can be attributed to high rates of strain specificity. We need to consider several factors like host range specificity, transient resistance, and strain sensitivity of the bacteriophages to understand the application of these bacteriophages as biocontrol agents or biopreservatives for sprouts.

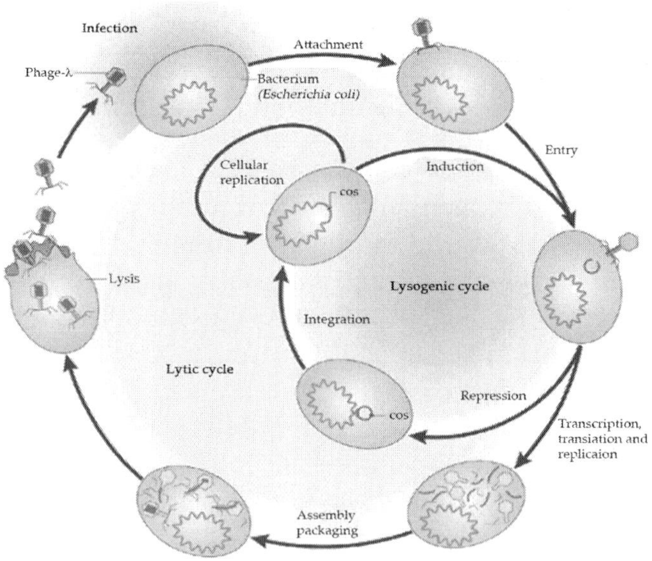

Figure 2.7. *Lytic bacteriophages.*

Source: https://www2.le.ac.uk/projects/vgec/highereducation/topics/microbial-genetics-1/bacteriophage.

Bacteriophages can survive both on controlled seeds and treated seeds; this will reflect that there are many presents in this sprouting process. The bacteriocins can be prepared from the culture broth concentration technique. The cultured broth will be dominated by producer strains. We can use the concentrated bacteriocins in the form of partially purified preparations in the food preservation industries. We can use pediocin and nisin as biopreservatives for the food products. We can also add natural preservatives. The immobilization of bacteriocins carry several advantages. We can easily apply immobilized bacteria on the food substrates. There will be minimal inactivation of bacteriocins as the interaction between enzymes, food components and bacteriocins is limited.

There will be minimal cross-contamination in this process. Pediocin when combined with citric acid can exhibit antilisterial activity. Nisin when added with phytic acid can also exhibit the same effect. We can combine all these compounds to store mung bean sprouts. Combined treatment can result in reduced local microflora of the food products. *E. faecium* can produce enterocins compounds, and they can exhibit antimicrobial properties over a broad spectrum.

Bacteriocins are being investigated with the different food systems to bring about the relation between the pathogenic bacteria damage and application of bacteriocins to regulate them. The enterocins can be produced in a semi-synthetic media based on the bi-products produced from cheese producing industries.

The effect of bacteria can be linked with the concentration and temperature of the samples being considered for studies. Bacteriocins were applied on green asparagus to know these compounds in the form of washing treatments. Most of the food products support complex ecosystems where the dominant microbial populations exist.

2.5. FISH PRODUCTS

Fishery products have attracted increased attention from customers that are thriving to maintain a healthy lifestyle. This is due to the fact that the fishes affect human health in many positive ways and can also provide maximum nutritional value (Beatriz et al., 2016). The nutrient composition in the fish products can harbor several foodborne pathogens and hence can be prone to spoilage. According to an estimate, at least a quarter of the total food products are spoiled due to microbe-induced spoilage. Most of the microbe-induced damage can be due at the postharvest stage. The lactic acid bacteria can produce hydrogen peroxide. Nicotinamide adenine dinucleotide and flavoprotein oxidase in lactobacillus can generate hydrogen peroxide. The sulfhydryl groups can be oxidized due to the action of hydrogen peroxide and this will denature several enzymes that are essential for microbial activity.

The membrane lipids can be subjected to peroxidation and this will increase the permeability of the membranes (Hans et al., 1995). *S. aureus* and *Pseudomonas* species can produce several problems within the food products, and they can be controlled by hydrogen peroxide. Most of the lactic acid bacteria that are heterofermentative in nature can generate carbon dioxide. This gas can bring about anaerobic environments, and this will drastically lower the decarboxylation of enzymes. Due to deposition of carbon dioxide in the lipid bilayer of the membranes, the permeability of the bacterial species will be impaired.

Most of the microbes that can spoil the food products can be inhibited by carbon dioxide. The psychrotrophic bacteria that belong to gram-negative group can be regulated by the application of carbon dioxide. Due to citrate fermentation, there will be production of aromatic component called as diacetyl. This product will be produced by several heterofermentative bacteria. Lactate will be produced in this process too. We can use diacetyl in flavoring milk, and this compound is having high value in the dairy industries (Sofia et al., 2019).

This compound can regulate microbial growth. This compound can act on gram positive bacteria, molds, yeasts, and gram-negative bacteria. This compound can react with Arg-binding proteins. Due to this binding capability, this compound can interfere with the mechanisms that participate with the absorption of the amino acids. *Lactococcus, Enterococcus, Lactobacillus,* and *Carnobacterium* can be observed in the fresh water and seawater fishes. We can add 2.50% lactic acid, 0.25% lactic acid culture, 0.25% potassium sorbate and 0.50% sodium acetate to control gram-negative bacteria. Using these compounds has controlled the odor of the fishes that are being stored for 13 days. Several colorants were used for the food products. Some of them are betalains, anthocyanins, chlorophylls, and carotenoids. These pigments can be deemed as safe as there are negligible effects on human health. The colorants can boost the immune system and can reduce the risk of cancer, obesity, and diabetes. We can add plant derivatives in order to reap bioactive advantages. They can promote health, enhance quality, can act as antimicrobial agents, antioxidants and can be used as colorizing agents (Elotmani et al., 2004).

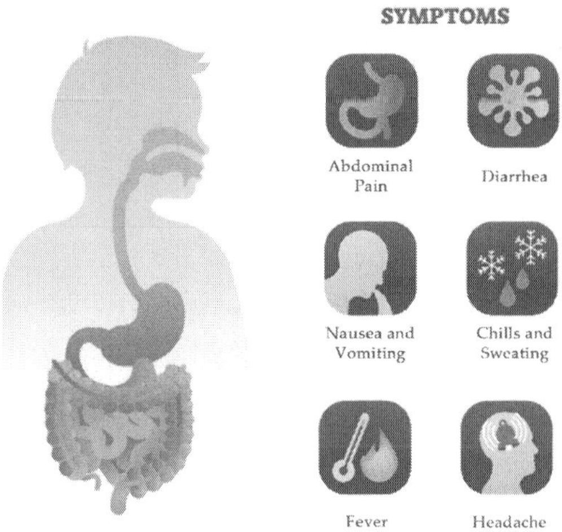

Figure 2.8. Foodborne illness.

Source: http://foodsafetytrainingcertification.com/food-safety-news/foodborne-illness-what-you-need-to-know/.

Foodborne illness (Figure 2.7) can be deemed as a major problem for human health in most of the countries. This problem is higher in developing countries than developed countries. Natural antimicrobials were prepared to reduce the growth of fungal population and bacterial populations. These antimicrobials have enhanced the shelf-life and quality of the food products. The synthetic preservatives have inflicted several health concerns over the last decade. The natural antimicrobials have become an alternative for synthetic preservatives since decades, and its popularity is being increased daily. These compounds were extracted from a variety of sources. We can obtain natural antimicrobials from fungi, algae, bacteria, animals, and plants. There are many studies that support antimicrobials that are produced from plants. The polyphenolic compounds that were produced from the plants exhibit increased chemical diversity and structural diversity. These compounds can bring about antimicrobial actions on pathogenic microbes. These compounds are hydrophobic in nature. Some of the essential oils can be classified as hydrophobic components (Lucy et al., 2006). These compounds can be produced from mushrooms and algae. The synthetic compounds that can be used in the preservation of the foods can be replaced by natural antimicrobial agents. The natural microbials satisfy the food safety requirements. Using natural agents can safeguard the native microbiomes of the consumers. The natural antimicrobials can damage the ATP molecules and they can disturb the proton motive force. Most of the mechanisms that are linked with the nucleic acids will be impaired. These antimicrobials will also disrupt the cell membranes of most of the pathogens. Most of the spices and herbs can act as antimicrobials and they will be produced from plants. Onion and garlic can inhibit microbes that are assumed to produce problems with human health. Vapors and juices that were extracted from the pants can inhibit microbes like *Staphylococcus aureus, Salmonella, Lactobacillus, Escherichia coli, Clostridium botulinum,* and *Bacillus cereus.* Some other fungal members that can be controlled by these compounds are *Saccharomyces,* Cand*ida* and *Aspergillus* spp.

2.6. OILS

The oils that are produced from onions can diminish the production of ethanol that is being produced by *Saccharomyces cerevisiae* and *Hansenula anomala*. These essential oils produced from onions can suppress the sporulation process in *Lodderomyces elongisporus*. The antioxidant capacity of these essential oils can impose antimicrobial action. The organic compounds like vanillin, citral, etc., were used in the food industries because of their antioxidant and antimicrobial activities. The essential oils from basil, lemon, oregano, rosemary, sage, and thyme can be used in combinations to control *Pseudomonas aeruginosa, Listeria monocytogenes,* and *E. coli*. The essential oil produced from oregano can act on *B. cereus*. The essential oils produced from thyme and basil can be combined to suppress the growth of the *P. aeruginosa, E. coli* and *B. cereus species.* Garlic can inhibit the growth of *S. aureus*. This includes thermonuclease and enterotoxin production. We can use spices that were extracted from fruits, leaves, seeds, and roots of the aromatic plants. All these components can act as flavoring agents. We can use rosemary, clove, cinnamon, and oregano to regulate the activity of the microbes. The cinnamic aldehydes and eugenol can be classified as major chemical compounds present in the cinnamon and cloves. The cinnamic aldehyde and cinnamon can suppress several bacterial members and fungal species. Some of the fungal species that can be regulated are *Saccharomyces, Penicillium,* Cand*ida,* and *Aspergillus species*. Some of the bacterial species that can be controlled are *S. aureus, Shigella, Salmonella, Listeria monocytogenes, Lactobacillus,* and *E. coli*. Eugenol and clove can inhibit fungal species and bacteria thereof. We can mix sodium benzoate or potassium sorbate with the cinnamon extracts to control *E. coli*. Apple juice can be added with these compounds to yield synergistic effects. Mustard, garlic, clove, and cinnamon can exhibit high inhibitory activity whereas oregano exhibit minimal activity on fungal species. *A. flavus* can be classified as a resistant microbe that can withstand the action of these essential oils. *P. roqueforti* can be classified as a sensitive species that can get killed on application of extracts from the spices (Ayala et al., 2007).

The thyme and oregano can deactivate or kill several microbes, and this can be attributed to thymol and carvacrol. These compounds can act on *Yersinia enterocolitica, Vibrio parahemolyticus, Staphylococcus aureus, Shigella, Salmonella, Proteus, Pseudomonas, Pediococcus cerevisiae, Listeria monocytogenes, Lactobacillus plantarum, Enterobacter fecalis, Escherichia coli, Campylobacter jejuni, Brochothrix thermosphacta, B. cereus,* and *Aeromonas spp.* These compounds also act on both yeasts and molds. Some of the organisms that can be controlled are *Schizosaccharomyces pombe, Saccharomyces cerevisiae, Rhodotorula, Pichia, Penicillium, Geotrichum, Candida,* and *Aspergillus species.* The antibacterial activity of these essential oils was tested along with the emulsifiers and stabilizers. Lecithin was used as an emulsifier and agar can be used as a stabilizer. The activity was tested at diverse temperatures (Burt and Reinders., 2003). It was obvious that thyme and oregano based essential oils are active against these microbes that can spoil food products. Rosemary along with camphor, camphene, pinene, and borneol can inflict high rates of antimicrobial activity. Sage with thujone can enhance antimicrobial activity. Most of these compounds kill bacterial species rather than fungal species. The extracts produced from sweet basil can act on both penicillium and mucor species, but they may not act on bacterial species. Many compounds from sweet basil in the form of essential oils exhibited antimicrobial activity on molds, yeasts, bacteria, gram negative and gram-positive bacteria (Delaquis et al., 2002). Vanillic acid and vanillin can be used against *L. Seeliger, L. grayi, L. innocua* and *Listeria monocytogenes.* Vanillic acid can exhibit severe lethal activity if the pH values are less. The essential oil like cilantro can affect the growth of fungal members. These oils will be produced from tea tree oil, savory, lemongrass, fingerroot, and coriander. Cruciferae members can produce antimicrobial compounds that can regulate the growth of the microbes. Some of the important members that can generate essential oils are rutabaga, turnips, mustard, kohlrabi, kale, horseradish, brussels sprouts, broccoli, cauliflower, and cabbage. These plant members can produce several antimicrobial agents that can kill bacterial species like *Pseudomonas, Lactobacillus's sake, Serratia, S. aureus, Salmonella* and

Listeria monocytogenes. The fungal members and yeast, including *Botrytis cinerea, Aspergillus flavus,* and *Penicillium expansum*, can be controlled by the essential oils generated from Cruciferae members. The action of these essential oils was tested in case of pre-cooked roast beef. The essential oil produced from horseradish was effective on the pre-cooked roast beef. The members of *Enterobacteriaceae* and *Pseudomonas* spp. were killed on exposure to this oil whereas lactic acid bacteria remained resistant and tolerant. The horseradish oil preserved the color of the cooked meat that is stored. These oils inhibited *Serratia grimesii, Listeria monocytogenes, S. typhimurium, E. coli,* and *S. aureus* in aerobic conditions at 12°C for a week. Phenolic compounds were also classified as important antimicrobial agents in the food industries. The triphenols, diphenols and monophenols can be called as phenolic compounds. We can use wood smoke in the preservation of foods. Some of the simple phenols that can bring about excellent flavor and odor to the food products are gallic acid, hydroquinone, and cresol. We can use liquid smoke to inhibit the growth of *Penicillium roqueforti* (Figure 2.8), *Penicillium camemberti* (Figure 2.9), and *Aspergillus oryzae* (Figure 2.10).

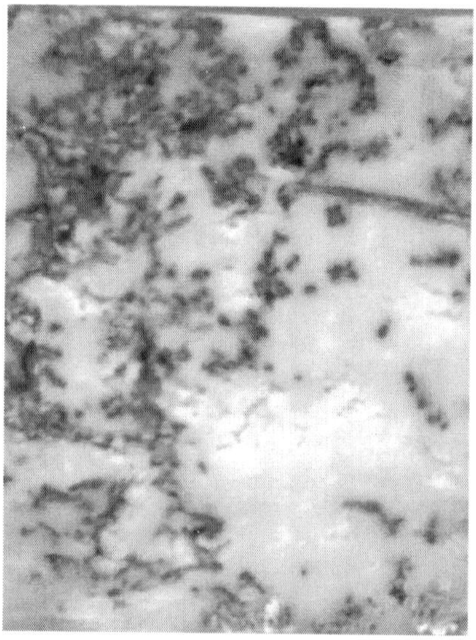

Figure 2.9. Penicillium roqueforti.

Source: https://upload.wikimedia.org/wikipedia/commons/f/f6/Blue_Stilton_ Penicillium.jpg.

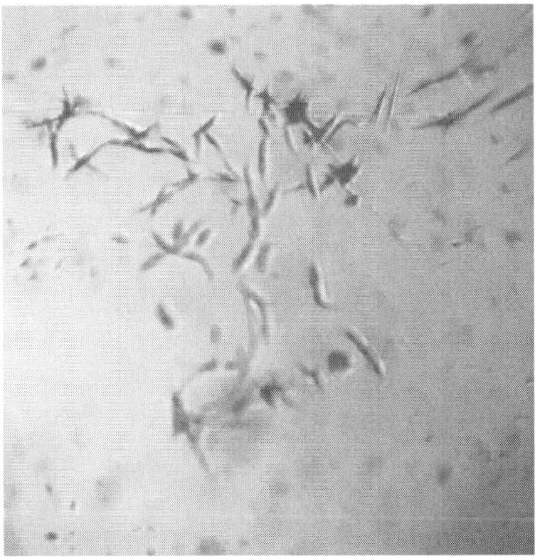

Figure 2.10. Penicillium camemberti.

Source: https://microcosmos.foldscope.com/?p=150984.

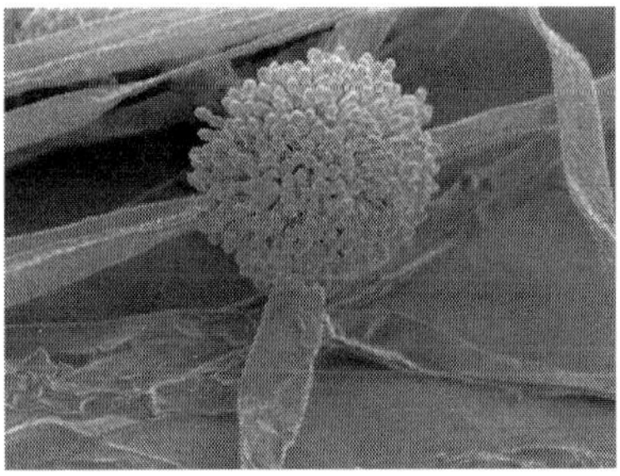

Figure 2.11. Aspergillus oryzae.

Source: https://sites.google.com/site/biol250aoryzae/.

It was reported that isoeugenol can reduce the growth rates of molds. Both p-cresol and m-cresol can diminish the growth of P. camemberti. A. oryzae growth can be inhibited by p-cresol, m-cresol, 4-methyl guaiacol and guaiacol. The phenolic acids can be extracted from plants, and these extracts can diminish the growth of S. aureus, Listeria monocytogenes, Salmonella, E. fecalis, E. coli, and Aeromonas hydrophila. MIC of about 64 micrograms per milliliter was observed when tertiary butylhydroquinone was used (Payne et al., 1989). Sinapic acids, ferulic acid, coumaric acid, and caffeic acid were classified as hydroxycinnamic acids, and they can be used to regulate the growth of E. coli, P. fluorescens, S. aureus and B. cereus. The hydro cinnamic acids can reduce the production of aflatoxins that are produced by A. parasiticus and A. flavus (Kim et al., 2004). E. coli can be killed by the addition of p-coumaric acid, ferulic and caffeic acid into the food products. Bacillus cereus, Staphylococcus aureus, and Escherichia coli can be controlled by p-coumaric acids. The parsnips, parsley, citrus fruits, celery, and carrots can have high amounts of furocoumarins, and they can inhibit the microbes that are present in the food products. The glycosides of flavonols, flavones, and catechins can be present in the strawberries, sorghum, plums, grapes, barley, and apples. These flavonoids can exhibit both antibacterial, antiviral, and antifungal activity. Quercetin can inhibit DNA gyrase. The resin produced from hop flowers can be used in the brewing industries. This will give good flavor to the beer products. Hop plants can produce xanthohumol and acylphloroglucinols that are prenylated. These compounds can inhibit gram positive bacteria. The acids produced by hop can inhibit fungi and protozoa. Isoxanthohumol and xanthohumol can act against Mucor rouxianus, T. rubrum, Trichophyton mentagrophytes, Fusarium oxysporum, and Candida albicans.

Chapter 3

Food Preservation

CONTENTS

3.1. INTRODUCTION

Any action that is made to maintain food to withhold the desired properties or its native properties for extended periods of time can be called food preservation. The art of food preservation has evolved into interdisciplinary science. The foods can be formulated, processed or raw in nature and some of the foods can be consumed by animals and humans for satisfying social demands, pleasure, inherent satisfaction, health, and growth. There are certain limitations that hold us back in consuming excess food consumption as it leads to several health complications. Some foods that constitute excess sugar, fat and salt must be avoided in all cases as they can produce several health problems in humans. The foods are primarily composed of carbohydrates, fats and lipids and a limited proportion of organic compounds and minerals. We can observe flavor producing compounds, polyphenols, pigments, antioxidants, acids, emulsifiers, and vitamins along with minerals in food products. We can classify foods into medical foods, functional, synthetic, harvested, nonperishable and perishable foods. We can classify the preservation methods based on the food products that might need to be formulated or prepared. To preserve foods, we need to understand the whole food chain that comprises stages starting from crop growth to the distribution stage (Lothar, 2000).

Figure 3.1. Food preservation.

Source: https://www.britannica.com/topic/food-preservation.

3.2. PRESERVATIVES

Nitrites and nitrates can be used in the preservation of food in the form of functional ingredients and preservatives. Food preservation (Figure 3.1) takes into consideration certain methods that can help us to preserve food with all the desired properties according to customer demands. Innovation is always important in technological and engineering developments. Food safety is the top priority in most of the food preservation industries that attempt sustainability and innovation. Preservation of food may not be classified as a straightforward process. There are several novel preservation techniques that were built or framed to bring about all the customer demands and satisfaction. Most of the preservation methods are initiated with complete analysis and steps that are to be followed in every stage starting from harvesting (Parke et al., 1992). The primary reason behind food preservation is to uphold the agriculture industry that may suffer due to inadequate planning and to provide certain healthy variations in dietary requirements. It is important to know that the agriculture industry provides several raw materials that can be supplied to diverse industrial sectors. The main aim of food preservation is to store food for extended periods of time without losing quality. Value-added food products are good quality food products that involve sensory properties, convenience functional and nutritional properties. The extent of food preservation can be based on consumer demand for food with extended shelf life. People like to consume foods with a variety of flavors and tastes. It is always essential to maintain diversity in food products. This is commonly seen in developing countries. There are certain important aspects that are to be considered in the food preservation process. They are the preservation time and the desired quality of food along with this source from where the products are being produced. After a certain period of time that preserves food, it may lose quality, and so it becomes undesirable. Quality considered as the changing concept. It is primarily based on satisfaction levels that customers put forward. If any food product transforms into an unsuitable product for human consumption, then we can call it a food product that has reached the end of shelf life. While studying the shelf life of food products, it is important to understand the rate at which the desired quality is being modified. Safety can be considered as an important attribute that must be considered as the first set of testing parameters. Second important attribute is the quality. Quality attributes can be variable; it constitutes microbial characteristics, sensory characteristics, and appearance (Maria et al., 2016). The quality might be lost easily and is mostly dependent on formulation composition storage conditions, packaging and consumption levels. If the preservation rates are decreased, then we can expect food spoilage that can lead to harmful effects on human beings. Preservation period is always essential in most of the food processing industries. It is important to know that after a certain period of time, at least one quality attribute of food may get lapsed resulting in an undesirable state of food product. This period will be determined by the manufacturer based on certain criteria well before sealing food products. It is always good practice to establish a good margin of shelf life. This will ensure that the customers do not get any health problems even after consuming food products after the mentioned shelf life on the package. If preservation fails, then we can expect the minor spoilage leading to color loss and it can also lead to hazard. There are several stages in the food production process that includes harvesting, storage, processing, packaging storage, distribution, sale, transport storage and consumption (Eklund, 1980). Food processor involves storage of fresh foods and food ingredients. The product quality can be defined by several parameters that includes microbial characteristics eating the characteristics field and appearance. Due to microbial growth, there will be degradation of food products that can be categorized as loss of color, and hence the physical framework of the food products will collapse, and this can be due to bruising in the vibration process. Half flavor is another important quality loss that can occur and this can be attributed to enzymatic reactions that can change the flavor. Due to mechanical problems, there will be cracking and problems with controlled release. Due to microbiological interventions, there will be production of toxins that can lead to altered flavors. Further, there will be rancidity due to redox reactions, loss of nutrients and nonenzymatic browning in the food products, if they are not professionally managed. There will be string cage recrystallization change in the phase and encapsulation of flavor due to physical improper management. The quality of food will be based on certain levels of parameters like occurrence of radioactive hazards, toxic chemicals, parasites, microbial toxins, pathogens, and level of microflora. The choice of customers is also an essential criterion to determine the quality of food products. This constitutes expectation attitude past experience media reportage of consumer, economic factors advertisement and public knowledge of the food product. There are several reasons that explain the spoilage of food. Spoilage and deterioration of food products can be due to microbial effects, physical and chemical effects along with mechanical problems. The primary factor that determines the food spoilage starts from improper handling of food products during all the stages of food processing

(Elena et al., 2017). This will result in a decreased shelf life that is highly undesirable. Minor cuts in vegetables and fruits while harvesting and even post-harvesting can lead to the production of fungal based infections. The tuberous vegetables get dehydrated if subjected to environments that are having low rates of humidity. There will be the possibility of wilting if agricultural products are subjected to an environment that may dehydrate their cells. If the dry foods are subjected to high humidity levels, they will become soggy, and it is highly undesirable in the food processing sector. Important microbial contaminants are from animal waste in air, water, and soil particles. The microorganisms such as protozoa, bacteriophages, bacterial, and fungal members can spoil foods. The mites and insects such as houseflies, fruit fly and others can also cause several problems in the food products. Rodents can also spread diseases if the foods are not properly preserved. Milk, fish, and meat can be considered as perishable products with a storage life of at least two days. Fruits and vegetables can be classified as semi perishable products with a storage life of at least two weeks, and this may vary according to the exposure time (Lynda et al., 2018).

3.3. FOOD PRESERVATION METHODS

The food preservation methods can be classified into inhibition inactivation to avoid recontamination techniques. In inhibition techniques we can use low temperature storage, hurdle technology, phase transition changes, removal of gas, chemical modifications Structural modifications, surface coating concentration drying, freezing, pH control, antioxidants, preservatives, fermentation, acidification enhanced carbon dioxide use, and decrease in oxygen levels. Inactivation methods constitute magnetic field sound light extrusion frying, cooking, blanking, pressure treatment, electrifying, irradiation, pasteurization, and sterilization. Avoiding recontamination methods, include risk management and analysis ascetic processing, hygienic storage, hygienic processing, and packaging. Food preservation primarily starts during the harvesting stage that mainly contain several microbes brought about by water, soil and plants. Meat produced from slaughterhouses may also contain several microbes that can require advanced preservation techniques. Most of the preservation methods will be dependent on the source of foods that means whether they are from animal origin or plant origin. The post handling of the raw materials is very important because it demands effective control of temperature, gas composition and humidity along with several other environmental variables (Nalan, 2018). There are several physical treatments that can be considered in the preservation process. They are waxing, cleaning adequate treatments with the temperature, precooling and curing. Meat can be considered as edible food material for several mammals and birds and hands will be exposed to diverse microbial cultures. Mostly, the quality of postharvest of the products obtained from slaughterhouses is particularly important. It is important to ensure sanitation and cleaning in almost every step of the food preservation process. We can use chemical disinfectants to control and kill microorganisms that can spoil food materials at all the stages of food processing. We can preserve fish products using decontamination methods, electrical stimulation, and chilling techniques (Zengin et al., 2011). We can now use hot water rinsing along with the chlorination process to decontaminate fish products. The decontamination process can be done with the application of ozone chlorine dioxide, chlorine hydrogen peroxide, and phosphate. One can also opt for treatments. Some of the important pretreatment methods are chemical treatments, physical treatments, and blanking techniques can be used before the actual preservation process. The primary purpose of these pretreatments is to enhance the quality of the products and to increase the efficiency of the process pertaining to food preservation. Debris treatment and changing processing strategy has attracted attention in several food processing industries.

We can use quaternary ammonium compounds, trisodium phosphate, iodine, bromine, peroxyacetic acid, ozone, hydrogen peroxide, chlorine dioxide, and chlorine as important components in any food processing systems. It is worth noting that fumigants can act as moderate preservatives. These fumigants can be used to regulate insect growth in any food product. The most popular fumigant that can be used in the preservation of food products is methyl bromide. Methyl Bromide can damage atmospheric ozone, and hence its usage was restricted and prohibited in food processing industries. It is always important to search for environmentally friendly fumigation techniques to safeguard the environment (Maan et al., 2002).

3.4. INHIBITION METHODS

The inhibition methods mainly depend on controls related to the environment. We need to consider temperature controls that participate in specific processing methods like microstructure control. There are many intrinsic properties that are present in foods like pH value and water activity. The temperatures between 5°C to 60°C can be considered as an optimal temperature zone for several microbes. The temperatures that are well below 5°C can be classified as appropriate food preservation temperature. Chemicals have been used to preserve food products since centuries. There are different varieties of additives and chemicals that can be used in the preservation of food to regulate pH, and they also act as antioxidants and antimicrobials. The functionality of the food will be preserved through this chemical process. There are certain additives that can be prepared synthetically. They are tertiary butyl hydroquinone and others that are classified as phenolic antioxidants, some of which are extracted from natural resources. Vitamin E can be considered as the natural antioxidant that can preserve food products. Food additives must be classified as safe in order to use in the preservation of food products. So far, the safe consumption by humans some of the food preservative that are permitted by law include, sodium propionate, propylparaben, ethylparaben, methylparaben, sodium benzoate, sodium diacetate, benzoic acid, sorbic acid, lactic acid, citric acid, acetic acid, nitrides, and sulfites. There are several problems with organic acids that can be used as microbial inhibitors in the food products. Some of the microorganisms may become resistant to the conditions produced by organic acids. Some of the microbial strains mainly exhibit inherent variability. There are several microorganisms that can use organic acids as metabolic resources, especially for carbon. The use of organic acids must be least used in case of high microbial growth. Several plants can produce compounds that can exhibit excellent antimicrobial activity, and these are called as biopreservatives and green chemicals. The use of green chemicals has increased since last decade (Yi et al., 2010). There are several spices and herbs that can be used to control bacterial growth. Antimicrobial agents are mostly present in the foods of both vegetable and animal origin. Most of the spices and herbs are being used as flavoring agents. There are several essential oils that can be used to protect the food products for extended periods of time. Allyl isothiocyanate that is present in mustard can be used as a preservative. pH can be considered as an important regulating factor that defines microbial reactions in the food products. The foods that are preserved under 4.5 pH value can be classified as low risk foods. Most of the microbes need appropriate temperature, nutrients, and water for growth and reproduction. Benzoic acid sorbic acid and propanoic acid can be considered as the best preservatives that can inhibit microbial growth effectively. Inorganic acid-like sulfate can be recommended in case of reduced pH values. The strong acids may not readily get penetrated into the cell membrane of the microorganisms. The acids that are supplied to the food products mainly lower the pH of the environment, and hence there will be enzymatic deactivation of all the cell surface (Hua et al., 2017). This will also reduce the cytoplasmic pH, and it can be attributed to enhanced permeability of protons. Weak acids generally penetrate the membranes easily as most of the membranes are lipophilic in nature. The pH of cytoplasm will be lowered due to the addition of weak acids. The acids that are not associated can impose several metabolic problems in the microbes. The ions like nitrate sulfate and carbonate can work as inhibitors at acidic pH. pH can be considered as an essential factor that regulates microbial growth, and it participates in inducing stability of vitamins and proteins formation of gel and inducing stability of enzymes. Some of the enzymes that participate in the lysis of microbial cells are being used in the food industry to break down several polysaccharides and to produce the latest food preservatives that have more advantages. It is to be mentioned that antibiotic resistance can create a lot of problems if we use chemicals without a proper monitoring framework. There are several physical modifications that occur in food products that are subjected to preservation. Water can be considered as an important component in most of the food products. The activity of water can be associated with the disturbance of food stability due to increased growth rates of microbes. The total amount of water that is present in the food materials then dictate the shelf life of food products that are being preserved. Minimum water activity can be defined as the limit where most of the microbes cease to reproduce. Generally, the limit of water activity may not exceed 0.7. Most of the pathogenic bacteria grow at 0.86 water activity. The molds and yeast may withstand fluid water activity, and it is to be mentioned that this species may not grow at a water activity of 0.62. Some of the Other factors that regulate water activity temperature heat treatment antimicrobial agents salt and pH. If we remove water from the food products followed by the addition of solutes, then we can expect the minimal water activity of that specific food product under investigation (Hye et al., 2019). Drying can be classified as a conventional method for food preservation. Most of the drawing process was done under sunlight in the older days. The liquid condition of the food products will remain intact even though considerable amounts of water is removed from the food products. The drying process can come with certain limitations, such as loss of functional compounds aroma and flavor. The cost of packaging of this dried food products is minimum when compared with transport of frozen foods and canned foods. If enzymes are used in the preservation process, it is always essential to ensure that the activity of these enzymes is maintained during the entire process of preservation. There are several antimicrobial enzymes that are hydrolytic in nature that contain essential components of the microorganisms such as fungi and bacteria. There are certain oxidoreductases that can work as antimicrobial agents and filter the reactive molecules that are being generated in the food products. Nisin and natamycin can be produced synthetically and through microorganisms. Even a lower concentration of these chemicals can decrease the microbial growth. Some of the microorganisms that are present in food become resistant and get populated in humans and animals. The liquid foods can be concentrated through membrane separation, freeze concentration, and thermal evaporation. The freezing process can change the physical state of the food products by transforming water molecules into ice as the energy is removed when subjected to freezing temperatures. Temperatures below −18°C can be considered as appropriate temperature if you want to store food for extended periods of time. Below −18°C most of the microorganisms can grow, but there will be certain nonenzymatic and enzymatic changes. This change occurs at a slow pace, and this will gradually change the organoleptic quality of the food products. The freezing process has become more popular than the drying process as the freshness of the food will be restored easily (Imram et al., 2015).

3.5. CLASSIFICATION OF FOODS

Foods can be classified as stable products in their glassy stage. During this stage, there will be stagnation of molecular motion, and most of the chemical reactions will be arrested. There are several theories to support the glass concept of foods. Edible coatings are being used in food systems in several industries all over the world. These coatings can improve the texture and appearance but the food products and ensure that there is limited water loss. Waxing of frozen fish and apples can be considered as example of edible coatings. We can apply mineral salt preservatives, fungicides, acidulants, and antioxidants in the process of surface treatments of foods. Pectin agar casein gum and jewels can be used as edible coating and the pressure functionalities and the structure of the food products for extended periods of time.

Most of the packaging techniques were based on manipulated gas compositions for decades. Due to respiratory activity that is found in plant products, there will be increased production of carbon dioxide and limited production of oxygen. This can slow down the ripening process. There are several techniques that can be used to maintain a modified atmosphere.

Type of technique, the composition of gas can be dependent on the purpose and the product that is to be preserved. The inactivation method uses heat energy. Increased amounts of heat will be used in the inactivation process. In the food preservation industry, thermal inactivation can be considered an important technique to deactivate several microorganisms. There are several advantages to thermal inactivation methods. The processed foods that are stored in sterile containers using thermal inactivation method can be preserved for extended periods. Most of the microbes that can spoil food products are sensitive to thermal inactivation processes, and most of them are heat sensitive. Taste and cooked flavor can be brought about by thermal inactivation process. Heat can be considered as a chemical free and safe process as far as food processing industries are concerned. There are certain disadvantages with the heat inactivation techniques. There will be a reduction in nutrients and due to over cooking, there will be disintegration of texture and undesirable flavors that may exist in the food products post-heat inactivation. We can use ultrasound and high pressure also in the food preservation process. There is always demand for fresh fruits with high quality, and in this case, people expect food that is subjected to less heat treatments and with minimal additives. Hydrostatic technology with high pressure can be used as an alternative for nonthermal treatment in the preservation process. Due to high pressure, there will be phase transition, conformational change, and other reactions that might occur in the food products.

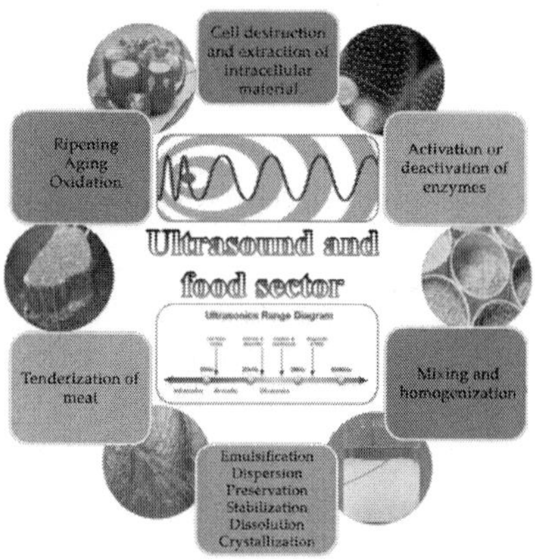

Figure 3.2. Ultrasound.

Source: https://www.mdpi.com/2304-8158/7/10/164#.

3.6. ULTRASOUND

Ultrasound (Figure 3.2) is another popular technique that can be used in the preservation process. It is nothing but sound energy that is equivalent to 20 kilohertz, and it is well beyond human hearing abilities. We can use ultrasounds directly in the food processing and food preservation process. We can categorize food materials based on ultrasound techniques. We can monitor food processing; we can test the quality attributes and can estimate the physical properties and chemical compositions of the food products. We can use electricity in the preservation process, and there are different forms of applications that use electric energy, namely high intensity pulsed electric field, High Voltage Arc discharge, low electric field stimulation, and microwave heating. Ionization radiation can react with materials by transferring a certain amount of energy (Aslihan et al., 2008). The electrons present in the material get ionized then resulting in the formation of positive and negative ions. This process mainly involves exposure of foods to the ionization radiation (Figure 3.3).

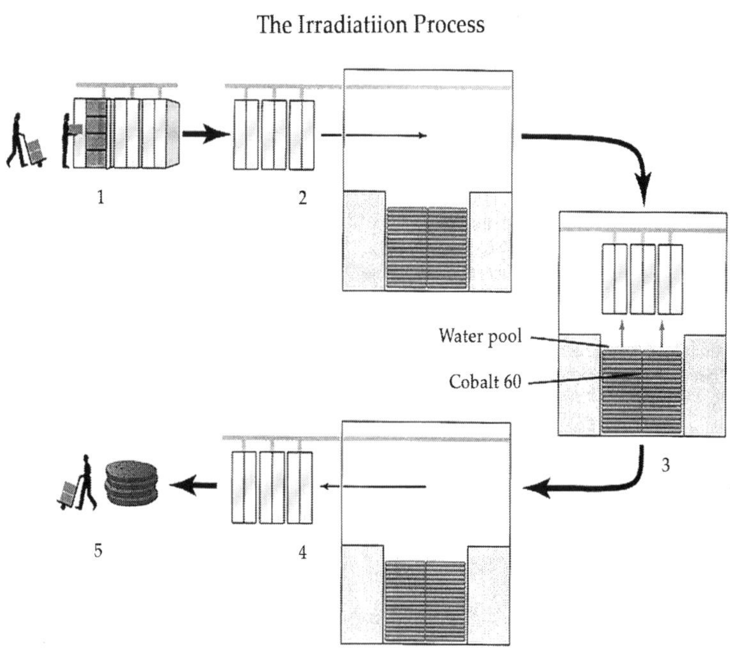

Figure 3.3. Ionization radiation.

Source: https://wiki.ubc.ca/File:FNH200_Lesson10_Facility.jpg.

The effect of radiation on biological systems is both direct and indirect, and it is the duty of the food processing unit to ensure radioactive guidelines and keep the exposure at a minimum. Given exposure to food materials through ionization radiation, there will be formation of hydrogen, hydrogen peroxide, hydrogen atom, hydrated electron, and hydroxyl radical. Hydrogen peroxide can be considered as a toxic agent for biological systems that can act as a strong oxidizing agent. Irradiation can increase the quality of the product, decontaminate and also extend the shelf life of food products.

Ultraviolet radiation can be used to kill bacteria in the food products. Ultraviolet radiation can bring about antibacterial activity using sunlight. We can use this radiation to sterilize thin liquid films and their molecules that generally have minimal depth. When used in higher concentrations, it can also damage the flavor and smell. The ultraviolet radiation can be considered as safe and ecofriendly, and it can be classified as a cost-effective technique that can be used by employees of chlorination systems. The photoreactivation process can also be used in the food preservation industries. If light is used to treat microbes, then there is the possibility that they can transform into sensitive species for visible light. This process can be called a photoreactivation process. UV light and white light can be used to damage bacteria and other species. Some of the Other species that can be damaged or killed by UV light and white light are cyst protozoa viruses spore and other fungal members. We can use pulse light as a sterilization technique to decontaminate volumes and surfaces. We can decontaminate medical products and pharmaceutics along with packaging materials using pulsed light. But we cannot easily differentiate inactivation and inhibition.

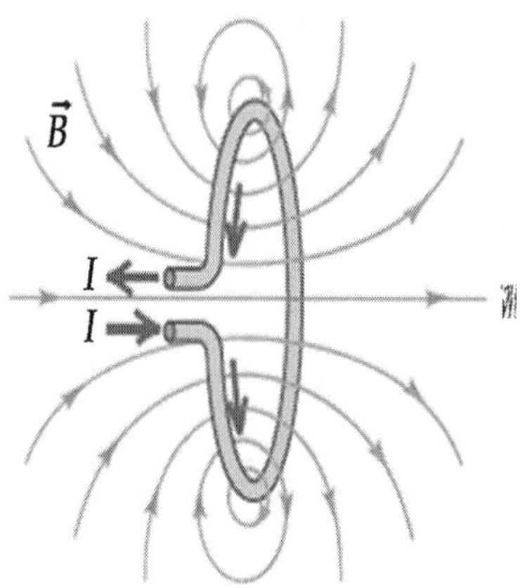

Figure 3.4. Magnetic field.

Source: https://encyclopedia.pub/2013

3.7. MAGNETIC FIELD

A magnetic field (Figure 3.4) can be used in the preservation process of food products. On application of magnetism, there will be attractive and repulsive forces on the respective materials. The magnetic fields can be used in sterilization and pasteurization processes. Avoiding recontamination is another important approach that can be classified as an indirect approach. The quality management tools and packaging tools can be used to avoid recontamination and contamination while preserving the process. Generally, these methods may not be classified as preservation techniques. There are certain standards that are framed by several agencies like hazard analysis and critical control point or HACCP, total quality management, hazard and operability studies, standard operating procedures, and good manufacturing practices. HACCP can be considered as an advanced method of protocols that ensures the highest grade of safety in the food production process. Most of the food production industries are starting to implement these protocols as they proved to be user-friendly. This system is based on a systematic approach, rational and scientific approach to control hazards, assess hazards, and identify them during all the stages of the food preservation process and production process. Mostly this concept is primarily based on two domains, namely prevention followed by bye documentation. This is a cost-effective approach. The World Health Organization has approved this system as it has several benefits. ISO 9000 is another important standard that is being followed by several food production and preservation industries that depend on consumer requirements. It is to be noted that ISO 9000 is a minimal standard for most of the companies that are involved in food production and preservation. This standard states that every company that deals with food products should comply with certain specific regulations provided by law. Good manufacturing practices must be adopted by food processing industries as they are considered as building blocks for the aforesaid protocols. Total quality management manual continuous assessment and improvement of the quality of all the processes that might occur in any organization. Hazard and operability studies can investigate problems associated with sequential stages in the production process of the food-based industries. Quality management alone may not provide food safety for the consumers, but it should be hazardous free and must be always safe. The food regulatory authorities should ensure strict rules so that all the food processing industries in that area can produce safe and hazard free foods. The combined methodology can also be called as hurdle Technology has attracted the attention of several food processing industries lately. In this technology, there will be usage of many preservative factors called as hurdles, and this proved to be an effective process in the food preservation sector. This can be called the hurdle effect. This will prevent the growth of microbes and at least slow down their growth in food products. There are several other preservation factors that are to be considered. Biotechnology has introduced several genetically modified goods that can provide value added products. These products can be easily stored, and their cost-effective nature attracts many entrepreneurs. Biotechnology can be used to enhance the functional quality of muscle foods, grains, vegetables, and fruits, and it also increases flavors in most of the foods. Using biotechnology, we can engineer several plants so that they can grow easily in different climatic zones. Xenobiotics in the food products can be investigated using diagnostic kits that were previously produced by biotechnology interventions. There are also turn on package sensors that were manufactured using biotechnology to investigate the extent of spoilage in food materials. It can also give information about the toxic products that might be released into the food products because of the microorganisms. Most of the food processing sector drives to reduce the extent of processing time and space. The entire process of food processing is an energy intensive process, and it requires a lot of expertise to match the industrial standards along with consumer satisfaction. There are always attempts to bring about effective methods and techniques that can reduce energy consumption during food processing and preservation. Food habits can be considered as an important component in any society since ages. The change in lifestyle and trends can result in demand for several attributes in the food products that industry should come up with. The functional and nutritional advances can also change the landscape of food processing industries. We need to focus on antinutritional factors with the help of systematic pretreatments where to preserve and processing stages. In most of the countries, integrated pest management is being practiced bringing about the best raw materials from the farm. The amount of pesticide residues has become a cause of concern in many areas that bye sex several sectors in our society.

Chapter 4

Post-Harvest Physiology

CONTENTS

4.1. INTRODUCTION

The postharvest period can be defined as a time where the food is separated from plant organs into the initial stages of processing at the industries then later can be prepared for human consumption. Vegetables and fruits are the living tissues that can be harvested at any stage of plant growth with enhanced moisture content and respective water activity. The harvested food products use water with respiration intact and can produce both water and heat using lipids, proteins, carbohydrates and other food Reserves. These are supplied by plants in the form of nutrients and photosynthates previously when they are attached with plant organs. Most of the vegetables and fruits can be consumed afresh without any processing. Some of the foods may be consumed in-process and forms like fermented products preserves dried foods, frozen and canned foods. The quality of raw material can also affect the quality of vegetable and fruit products that are being processed (Tadesse, 1991).

Figure 4.1. Postharvest technology.

Source: https://agriculturistmusa.com/postharvest-technology-definition-objective/#google_vignette.

4.2. FOOD QUALITY

Food quality is nothing but the amalgamation of properties, attributes, and characteristics of the food materials that can be consumed by humans. There are several quality attributes that are to be considered while handling raw materials. Some of them are the presence or absence of seeds, stiffness, and size of the raw materials. Some of the nutritional characteristics that are to be considered are functional components, antioxidants, and vitamins. There are certain sensory characteristics that are to be considered, namely smell, flavor, taste, texture and color. It is worth noting that there should be evaluation of quality from time to time by measuring the safety of the product's nutritive value, flavor, texture, taste, and color. We need to follow the safety aspects to preserve the quality attributes for extended periods of time. It is always important to understand the suitability of the raw materials that may arise from different sources and diverse climatic zones. We can also reduce spoilage by evaluating postharvest (Figure 4.1) quality of the plant products regularly. The quality of vegetables and fruits can be directly proportional or related to the native physiological status when they are present in the plant system (Dov, 2011).

4.3. ESSENTIAL FACTORS

There are several factors that regulate the quality of postharvest of vegetables and foods. They are preharvest factors harvesting factors physiological factors postharvest factors diseases insects' pest's quality c mechanical injuries. Preharvest factors mainly constituted cultural climatic and genetic factors. Harvesting factors involve a stress maturity harvesting method that is to be used. Physiological status constitutes disorders senescence ripening water loss and respiration. The quality factors that affect the postharvest quality are appearance, structure, flavor color and texture. Some of the postharvest factors are stress, light atmosphere composition, humidity, and storage temperature. Genetic factors are especially important as far as preharvest factors are concerned. Genetic constitution of any plants, animal derived raw materials can define the quality of Pre-harvest food product. The rootstock and cultivar selection can affect the storage potential quality and composition of the product. These important characteristics will be inherited by several genetic recombination's that might occur in the agricultural fields. Most of the fruit cultivars will be produced in the agricultural fields that are intervened by Limited amounts of pesticides and herbicides. It is worth noting that grape varieties that are used in the Fresh Food market can be different from grapes that are being used to produce wine. There are different sets of criteria's that are being followed by breeders in order to reduce new varieties that have enhanced processing features, minimal toxic compounds nutritional values, increased compositional values, disease resistance, and higher yield. Most of the food processing industries will directly purchase agricultural produce from the farmers based on their specifications during transit (Tiwari et al., 2013). There are several transgenic vegetables and fruits that have increased shelf life and minimal softening attendance is with limited browning sensibilities. Climatic factors also play a major role in the case of pre-harvest produce. The quality of vegetables and fruits will be affected by rainfall elevation, soil texture, wind light humidity and temperature. The harvest and its quality will also be affected by the quality of light intensity and duration of light during the cultivation period. The Shading of fruits due to leaves in case of tomato crops can result in an increase in its red color at the ripening stage. There are several transgenic fruits and vegetables that have improved quality features. Some of them are tomatoes, strawberries, potato peppers, cucumbers, eggplants, melon banana and apples. The crops that are grown in increased sunlight may contain high sugar concentrations and increased levels of dry matter. Citrus fruits that are grown in a shaded environment may contain more juice and acids than the crops that are grown under the direct sunlight. The crops that are grown under the sunlight for extended periods may result in high solid content but decreasing the number of acids and uses with thinner rind. The variations in light quality and length of the day will affect the physiology of fruits and vegetables (Jose et al., 2014). Onions produced in the zones where there is short day length mainly exhibit small bulbs. Dabble and wallet regions all the visible wavelength of light can result in the formation of anthocyanin pigments in eggplant and purple cabbage. Reproduction of thymine in most of the plants can be regulated by light, and this week, mint is mainly found in leaves of plants. The turnips that were grown in agricultural fields must be harvested during the early hours of a day such that the riboflavin concentrations will be higher during that period. The fruits that are grown in the soils exposed to cold climates may be acidic in nature mostly. Cultural practices are especially important and there to be considered well understanding preharvest factors. Some of the cultural practices are the density of planting chemical spray pest control thinning pruning water supply soil nutrients and soil type. All these factors can affect the quality of pre-harvest produce. The addition of fertilizers can influence the proportion of mental content that is to be present in the fruits, and the thinning and pruning process may also alter the nutritional components by altering the size of the crop load. If the planting used to close, then the sweetness of fruits will be drastically reduced. There are many physiological problems that may surface because of inappropriate soil management and poor nutrition status of the soils. Report items that were grown in low moisture soils or peat exhibited low dry matter than those grown in light loamy soils and Sandy soils. Due to the presence of high nitrogen and potassium ratio and reduced phosphorus in soil, the potatoes that are grown will become dark post-cooking. Due to irregular management of soil nutrients, pineapple suffers from opaqueness and undesirable flavor. Most of the pesticide residues can be observed in process ID products and fresh products. If the use of pesticides is increased, then there will be e production of harmful metabolites that may lead to several heart ailments in human beings. Harvesting factors are another crucial factor that determines the post-harvesting physiology of vegetables and fruits. It is essential to investigate whether the products that are being harvested are at maturity or not. The maturity levels of the harvest determine the quality and composition and result in dynamic storage potential. It is always important to maintain optimal harvest maturity and extend the life of postharvest products. It is important to harvest fruits when they are mature but did not reach the ripening stage to enable an effective postharvest handling process. The fruits that are ripened or over-ripened may become too soft and hence create a lot of problems in the preservation process. If the fruits are harvested too early e into the season, then we can expect several physiological disorders that may also practically lower the storage life. Maturity indices must be fallout while harvesting process. The maturity stage may not be suitable for making gems freezing and dehydrating, but they can be canned. Most of the fruits that are ripened can be used for drying, and we can also prepare concentrated food products like tomato sauce that can be used as supplements in several food products.

It is crucial to train the workers that are involved in the harvesting process. They made several contaminants in the agricultural produce. Quality control should be maintained by considering the established food regulations. All the pickers of vegetation and fruits must be trained such that they can harvest appropriate produce. There are several postharvest factors that ought to be considered. The fresh vegetables and fruits can contain excess water. Watermelon has at least 95% of water, and likewise, many fruits contain water, and they may use water into the atmosphere. Due to loss of water, there will be reduced to crispness, wilting, and shriveling. Because of excess loss of water, the tissue in the fruits will become mushy and tough. Due to this, the fruits and vegetables will become unacceptable by the consumers. Injury to plant tissues, periderm, lenticels, stomata, presence or absence of cuticle, surface type, and surface to volume ratio may determine the rate at which water will be lost. The loss of water can be reduced by suitable packaging, by eliminating injury to the products, increased pressure, minimal air moment, minimal temperature and increased atmospheric relative humidity. All these conditions must be maintained during storage and transport. In the case of vegetables, the optimal relative humidity must be maintained between 90% and 98%, whereas for fruits, the optimal relative humidity can range between 85% and 90%. Pumpkins and dry onions may require optimal relative humidity ranging between 70% and 75%. Some of the roots harvested from plant systems need to be stored at 100% relative humidity. Because of increased relative humidity, there will be several physiological disorders that may appear in the agricultural produce. Temperature is another essential factor that is to be considered in this process. The management of temperature can be classified as an essential tool in handling postharvest products, and this will define the pathological and physiological deteriorations. Based on the physicochemical properties of vegetables and fruits, the effective temperature may differ (Elena et al., 2016). We can expect enhanced shelf life with reducing temperatures in the case of green onion and lettuce. Atmospheric gas composition is another important factor that plays a critical role in the postharvest process. Respiration is an important physiological process that will be controlled by ethylene carbon dioxide and oxygen, and most of the microbial degradation is based on atmospheric gas composition. We can transport bananas with 5% carbon dioxide at 3% oxygen in order to avoid fast ripening. Light can also affect the quality of vegetables and fruits by regulating the degradation of pigments like keratin oils and chlorophyll. And this primarily occurs r due to oxidation of lipids. There will be production of toxins, degradation of vitamins like riboflavin and ascorbic acid and sprouting, if sufficient light is not supplied to the harvest. If the potatoes are exposed to light during the storage process, there will be formation of green tissues, and they are dominated by a toxin called solanine. It is essential that the intensity of light should be minimized according to the crop produce and harvest. We need to use Saturn packing materials that can prevent light and help in storing the harvest in dark conditions. Mechanical injury is the main result due to improper handling of Crop produce, and this will expose tissues to two microbes such that there will be contamination that may result in several physiological problems. Their reactions and chemical reactions, along with increased respiratory rate, decrease the overall quality of the food products. The flatness of the epidermal cells may be correlated with the strength to resist cracking during transport of tomatoes. The punches, cuts and surface cracks will affect the outer layers that are meant for production and result in loss of water. If the tomatoes are any other food vegetables are at initial stages of maturity, then there will be a chance of repair of the damaged tissues. Once the tissues are mature, then the chance of repair is much lowered. Some of the factors that affect mechanical injury are strain level, failure stress, area of contact, stiffness of tissue, impact surface, radius of curvature, mass of Agricultural produce, velocity, to impact and drop height.

4.4. DISEASES

There are several postharvest diseases that are to be considered. Some of the diseases are made at the initial stages of development and when intact with the plant body. Diseases and infections may also result due to protection of bacteria and fungi. The entrance of bacteria and fungi can be a true cuticle. In some cases, they may enter through natural openings and wounds that might appear on the surface of the agricultural produce. It is important to know that microbes can only attack damaged tissues. We can control postharvest diseases based on retardation capabilities of bactericides and fungicides. We can also restrict these diseases by preventing incipient infections.

The ontogeny of plants may affect the physical aspects of vegetables and fruits. The tissues that are at the initial stage of the development mainly exhibit shorter shelf life. Respiration is another factor that we need to consider understanding postharvest physiology. Most of the living species convert matter and Transformer matter into energy. All living things carry out the process of respiration. In this process, there is an enzymatic oxidation of substrates like organic acids, lipids proteins, and carbohydrates. The ratio of carbon dioxide that is produced after consumption of oxygen can be called the respiratory quotient. The lipids have respiratory quotients less than 1 whereas, organic acids exhibit respiratory quotients of greater than 1, and carbohydrates can exhibit RQ value of 1. In presence of anaerobic respiration, there will be increased RQ value. We can observe three different stages of respiratory Pathways. The first stage is to break down the polysaccharides into simple sugars as this process occurs through hydrolysis; the second step is to oxidize the simplified sugars into pyruvic acid; this process can be called glycolysis. The sugar will be converted into pyruvic acid, and there will be an aerobic transformation of this molecule into energy, water, and carbon dioxide. This conversion can take place through a citric acid cycle followed by an electron transport chain at the mitochondria. The production of ATP shows the respiration process in the case of vegetables and fruits is least effective. Most of the energy that is being generated by the breakdown of glucose will be released in the form of heat. Thermogenic respiration that might occur in vegetables and fruits can result in loss of polysaccharides.

It is important to know that the enhanced carbon dioxide concentrations along with lowered oxygen concentrations can diminish the respiration rates. The factors that can primarily control the rate of respiration are maturity level, age, cultivar type, produce type and temperature. It is worth noting that 5% of oxygen dip may not produce visible problems with the produce under study. The increased carbon dioxide levels can bring about severe damage to the vegetables and fruits. The oxygen threshold can be linked with the injuries that might form on the vegetables and fruits. The development stage of the harvest can also affect the respiration rates. The actively growing vegetables can exhibit high respiration rates to that of storage organs as seen in mature fruits and potatoes. The respiration rates can be directly proportional to the maturity of the harvest. The respiration rates can be affected by cultivation period, harvesting method and handling systems. The injury to the vegetables and fruits can enhance ethylene production that can also affect the respiration rates and hence the quality of produce will be poor. The water stress and transpiration levels can also affect the physiology of the postharvest produce.

Transpiration can be classified as an important reason behind water loss in the plant systems. The loss of water may lead to loss of nutritional quality, texture, appearance, and weight. The textural quality that can get affected can be observed in terms of juiciness, crispness, limpness, flaccidity and softening capabilities. Shriveling and wilting can affect the appearance of the product.

These processes may result due to plasmolysis. If the loss of weight falls above 5%, then the quality of the product will be affected and may lead to denial by the potential customers. Some of the factors that determine the transpiration rates are solute concentrations in the harvest, endodermic effects that might result due to prolonged evaporation, maturity level, heat produced due to respiration, movement of air, variations in the water pressure differential, surface area dimensions and structure of the skin.

The loss of water can be regulated by the protective covering at the external side of the produce. Mostly, transpiration occurs at cuticle, trichomes, lenticels, epidermal cells, stomata and hydathodes. Ripening and senescence can be considered as essential indicators that may define the quality of the postharvest produce. The eating quality can be regulated by the sensory attributes, texture, and color of the vegetables and fruits. Mostly, ripening may bring about the enzymatic, biochemical, chemical, physical and structural changes. There will be degradation of cell wall, hydrolysis of starch and cleavage of chlorophyll. There will be production of ethylene, aroma volatiles, anthocyanins, and carotenoids during the ripening stage of the produce. At the ripening stage, the endocarp will be lignified, epidermal hairs will be lost, cuticle gets thickened, and we can observe visible epicuticular wax.

We can also observe the texture loss, color variations, chloroplasts may get converted into chromoplasts, plastid changes, softening, enhanced permeability and cell wall thickening while ripening stage of the food produce. The loss of texture can be related with the change in the cell wall. Some of the cell wall changes that are to be considered are pectin depolymerization, pectin solubilization, loss of arabinan and galactan. Jasmonates, ammonia, sulfur, aldehydes, alcohols, carboxylic acids, and terpenes can affect the physiology of the fresh fruits and vegetables.

They can also act as insecticidal and antimicrobial agents. The atmospheric composition, cultivar, disease, and injury can bring about increased concentration of these compounds. The phytohormones can play a crucial role in the metabolic pathways of plants. They can control the development of plants in all the stages. Some of the processes that can be regulated by phytohormones are flowering, abscission, sprouting, rooting, dormancy, rest, and ripening.

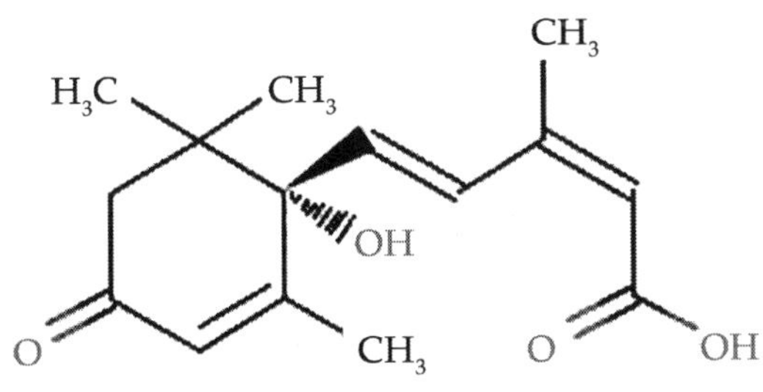

Figure 4.2. Abscisin.

Source: https://www.digopaul.com/english-word/abscisin.html.

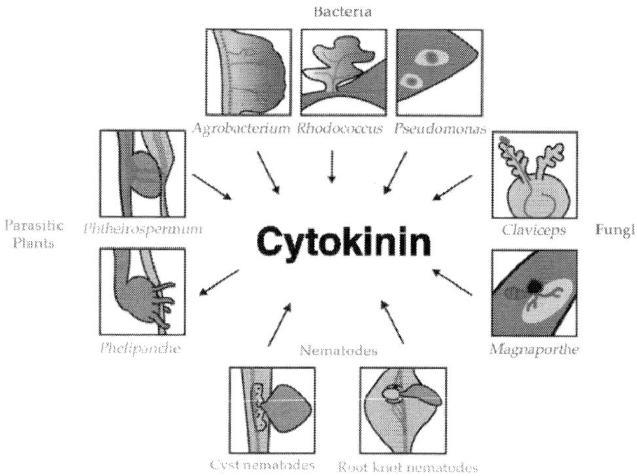

Figure 4.3. Cytokinins.

Source: https://ars.els-cdn.com/content/image/1-s2.0-S1369526617301966-fx1_lrg.jpg objective/#google_vignette.

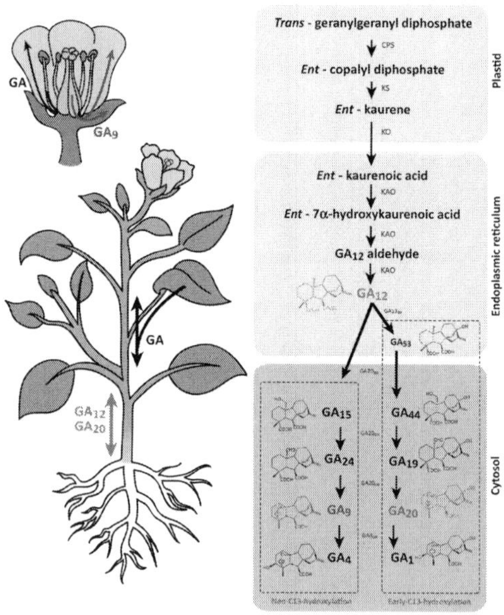

Figure 4.4. Gibberellin.

Source: https://marlin-prod.literatumonline.com/cms/attachment/287b1c78-c279-4ce7-8c05-f07be449f365/gr1_lrg.jpg.

The Abscisin (Figure 4.2), cytokinins (Figure 4.3), and gibberellin (Figure 4.4) can also affect these physiological processes to a greater extent. Most of thc postharvest physiology encircles with the dynamics of ethylene production. Ethylene can be produced from methionine in plants. In the first step, there will be formation of S-adenosyl-L-methionine or SAM from methionine. This step will be mediated by SAM hydrolase. This compound (SAM) will be converted into 1-aminocycloproane-1-carboxylic acid or ACC. The second step will be catalyzed by ACC synthase. The ACC that is formed in the second step will be converted into ethylene, and this conversion will be mediated by ACC oxidase. It is important to know that ethylene production in plants is dependent on the oxygen concentrations. The bioactivity of ethylene will be based on carbon dioxide and oxygen levels.

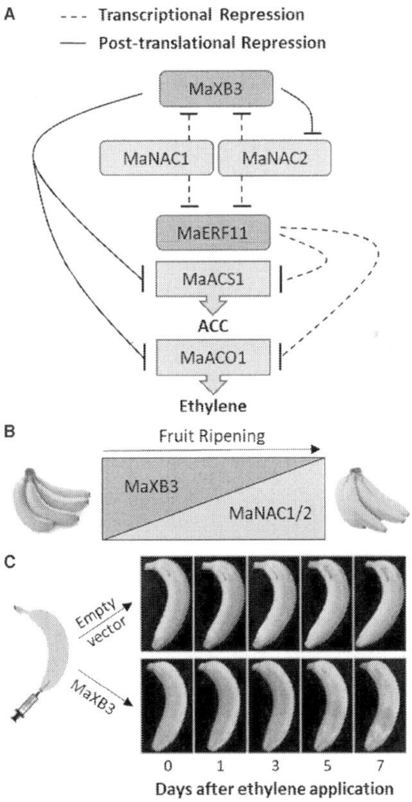

Figure 4.5. Ethylene.

Source: http://www.plantphysiol.org/content/184/2/568

Ethylene (Figure 4.5) can induce ripening of several climacteric and non-climacteric fruits. It also participates in the metabolism of phenylpropanoids, respiration and chlorophyll degradation. The physiological disorders can be due to metabolic disturbances in the plants. The external factors and nutritional imbalance can alter the physiology of the plants. Due to physiological disorders apples suffer with water core, bitter pit, brown heart, storage scald, and sunburn. Sunburn can be caused due to heat stress conditions. Storage scald can be caused due to injuries while chilling. This can be attributed to the production of conjugated trienes. These compounds can be produced from the oxidation of alpha farnesene compounds. The internal breakdown in the apples can be called brown heart. This can be due to injury caused by carbon dioxide injection. Bitter pit can be caused due to high potassium levels and low levels of calcium. This can be caused due to accumulation of the magnesium molecules. The middle lamellae will be dissolute due to action of succinic acid and oxalic acid. There will be altered permeability for potassium, and the proton secretion capabilities will be changed. Water core is another important physiological disorder that can be observed in the apples. Due to this disorder, the conversion of sorbitol into fructose will be stopped. This will result in deposition of sorbitol in excess amounts. In case of sunburn or sunscald, there will be yellow or white spots on the fruits that were exposed directly to the sunlight. Some of the areas that are exposed to sunlight will be sunken and dried out (Bleecker et al., 2000).

We can observe senescent scald, watery breakdown, internal breakdown, flesh spot decay and core breakdown in pear crops. The core breakdown can be differentiated due to soft breakdown of the core tissues, and there will be browning of the tissues. This condition exists due to the storage of the fruits for extended periods of time. Flesh spot decay can be demarcated by partial browning and we can observe cavities. In case of internal breakdown, we can observe dark brown flesh and reasons that might cause this disorder are still to be assessed. The watery deterioration can be observed in case of watery breakdown disorder and this can occur due to exposure to warm temperatures or gradual cooling. We can observe inappropriate odor and taste along with the black discoloration in case of senescent scald. This occurs when there is end of the postharvest life and the produce is getting prepared for fermentation. The pomegranates may suffer with chilling injury and husk scald. There will be loss of red color in most of the arils. We can observe browning of the skin color and husk will be discolored. This can be caused due to oxidation of phenolics, and this can occur due to exposure of these fruits to low temperatures. This is obvious if the temperatures fall below 5°C. Some of the stone fruits like prunes, plum, nectarine, peach and apricots may suffer with the internal breakdown. This can be attributed to loss of flavor and appearance of blackened pit cavity. There will be red pigmentation along with the browning of flesh. The amount of juiciness can be lowered. We can also observe leatheriness and mealiness in case of this disorder. This occurs mainly because of the fruits being exposed to the room temperature suddenly post-cold storage. This is more evident while ripening. This is worth noting to relate the mealiness effect with the genetic intrusions. The stone fruits can suffer with black stain and inking problems. This can be demarcated with the black and, in some cases, brown spots on the skin surface. This can be caused due to abrasion, and this can be attributed to the metal contamination with aluminum, copper, and iron. This crop can also suffer with bruising and surface pitting. This will result in the production of large, flattened areas with minute sunken regions. This can result due to compression and mechanical impact of the storage equipment. Tomatoes can have problems with physiological disorders like puffiness, catfacing, fruit cracks, sunscald, blossom end rot and blossom drop. There will be an empty cavity in the fruits, and we can also expect hollow inside of the tomatoes. This condition can be due to inappropriate pollination, and this can be attributed to the occurrence of heavy rains, increased nitrogen fertilization, and temperature extremes. The catfacing disorder can occur due to chilling and cloudy weather while blooming. This can be demarcated by scarring and puckering of the terminal portions of blossom. Fruit cracks can also be considered as an important problem as far as the postharvest physiology is concerned. This condition can be explained with the concentric and radial cracks in the fruit surface. This condition occurs due to irrigation and heavy rainfall after drought conditions. The sun scald condition can be represented by yellow spots and shiny white spots on the surface of the fruit that got exposed to the direct sunlight. There will be sunken areas and dried areas. This can be due to extreme heat generated during direct sunlight. The blossom end rot is another important condition that is of importance in the preservation of foods. We can observe black spots, and this can be attributed to the increased nitrogen fertilization. The plants might need minerals in perfect balance, and this defines the growth of the crop and so the harvest. The supply of balanced nutrients can be classified as one of the essential factors that completely regulates all the metabolic processes. The sensitivity of the tissues can be dependent on the water relations, nutrient transport, and growth of the fruits. Most of the physiological disorders are related with the balanced mineral input. There are several disorders that might result during the ripening stage and storage. The disorders can be induced due to storage factors, and in some cases, we can predetermine the fate of the produce even if it is in contact with the tree or plant. Most of the predetermined factors will be associated with the mature stages of the plant species. The enzymes like peroxidase, catalase and polyphenol oxidase can mediate oxidation of the phenolic compounds, and it results in browning of the tissue. Tissue softening can be caused due to activity of polygalacturonase. This enzyme can hydrolyze glycosidic bonds that are present in pectin, especially that are present in the polygalacturonic acid. The stiffness of the tissues can be due to activity of pectinesterase. This enzyme is mainly responsible for the hydrolysis of the ester bonds. Lipoxygenase can be involved in the lipid oxidation, and it can produce off-odors and off-flavors. There will be problems with the nutrition quality if the ascorbic acid oxidase is overactivated. It can oxidize ascorbic acid that can reduce the ascorbic acid concentrations in the fruits (Napier et al., 2006). The phytol ring present in the chlorophyll molecules will get cleaved due to the activity of chlorophyllase. This can result in impaired green pigmentation. Amylases can hydrolyze pectin and amylose. This can enhance sweetness and damage texture in the fruits. The protein hydrolysis will be due to proteases, and this will reduce the digestibility rates. Therefore, the nutritional quality of the fruits will be reduced. Hydrolytic rancidity can result due to lipid hydrolysis induced by lipase. The phosphates will be released due to hydrolysis of phytic acid that is mediated by phytase. Glucose oxidation will be mediated by glucose oxidase, and this will result in the production of hydrogen peroxide.

There are certain quality specifications that are to be considered in the case of processed vegetables and fruits. The fruits like tomato, apple, and citrus can be studied for quality specifications pertaining to flavor, sugar content and acidity. The fries and chips of taro, banana, and potato can be checked for reducing sugars, starch content, and texture. Pear, peach, and apple can be classified under canned products. They can be checked for flavor, texture, and color. Some of the preserves like peach and apple can be checked for acidity, pectin content and sugars. The pickles made up of cabbage, olive and cucumber can be checked for texture, sugar content, and composition. We need to investigate the total solids that might be present in the puree and sauce made by apple and tomatoes. The acidity and fermented sugars should be studied in the case of alcoholic beverages made up of apples and grapes. The solid content and composition of apricot and mango must be considered while examining the dried products. It is important to consider the flavor, texture, color, and composition of the frozen products that are made up of onion, carrot, and pea plants.

Packaging is another crucial factor that is to be considered while defining the quality of the food products. We need to consider the packaging sectors as a handling unit installed for an important purpose of food safety. There should be a home storage unit and warehouse unit. We need to reduce waste and the quality should always be protected. The marketing and transport costs should be minimized as much as possible to establish appropriate cost balance scenarios. We need to attempt the latest transportation modes that can lower the transaction costs. There are two primary packaging types that are small size packagers and large sized containers. We can opt for room cooling to store the food materials. The cold air will be supplied into the food materials to take away heat. Mostly, air will be cooled with the aid of refrigeration systems. The removal of heat will be easy if stacking and vents are in place. The surface of the container should be in contact with the cold air so that the products will be maintained at low temperatures. Package ice is another technique that can be used in preserving the postharvest products. We can use hydrocooling techniques in case the food products are to be cooled rapidly. In this technique there will be flooding or spraying of water into the product. The temperature can be close to freezing. Vacuum cooling is another technique that can be used in this case. The cooling effect can be due to evaporation, and this will take place from the products themselves (Paola et al., 2002). This evaporation process may take place in a vacuum in a chamber. There will be lowering of the vapor pressure. Evaporative cooling is another essential method that can be used in the preservation process. The precooling of tomatoes can be done with forced cooling of air. We can use package ice and the hydrocooling process in case of spinach. Hydrocooling, forced cold air and hydrocooling can be done in case of pear crop produce. For peach, we can opt for hydrocooling and forced air.

In the case of apples, asparagus, beans, broccoli, and others, we may choose hydrocooling technique. We need to ensure that the main objectives of storage are met. The muscle biology can be classified as a complex tissue that can involve several physiological functions and this dynamic activity can be considered while preserving meat products. The muscles play an important role in the protection and deposition of proteins. The slaughter conditions can be considered as important in the preservation technology. The meat products can be susceptible to several contaminants due to enriched ingredients in them. It is assumed that meat that is derived from healthy animals is free from microbes. The metal products can be contaminated by diverse slaughtering processes, intestinal contents, hooves, skin, polluted water, and air mostly. Most of the animal skins can be contaminated with microbes like molds, yeast, pseudomonads, micrococci, and staphylococci species. The slaughter facilities, utensils and equipment must be properly sanitized such that the microbial load will be reduced. The bacterial load will be affected by the type of animal species. The carcasses can be subjected to a minimum of 60°C in order to lower the bacterial contamination and to remove hair from the epidermis. The organic acids can be used to decontaminate the meat products efficiently. The effect of organic acids can be controlled by the acid molecules, dissociation extent and pH of the organic acids. The effectiveness of acids can be based on the slaughter time and exposure of the meat products to acid, microbial type that might exist on the surface, temperature and concentration of the acids, acid type, bacterial load, and the surface of meat that is being exposed to the microbes. The effect of acid decontamination will be increased with the rise in temperature of the lactic acid in the meat products. The decontamination of the meat can be proportional to the fat content of the meat products. Most of the bacteria may be attached to the surface of the meat. Bacteria retain on the liquid surface of meat products. They can also get trapped on the meat surface. This can be investigated by microtopography off the meat surfaces. In some cases, bacteria species can get absorbed into the solid-liquid interface that might exist in the interior layers of meat. Adhesion is another process that can be used by bacteria through polymer bridging. It is important to treat the surface of meat as quickly as possible post-dressing process, and this can also be done after the boning process so that there will be Limited contamination by bacteria. Bacteria can attach to the surface of meat in a two-step process that can be seen in most of the cases. There will be an Irreversible attachment and reversible Association in this process (Bunger et al., 2004). We can attempt a washing process to eliminate bacteria that might be hidden in the water film at the surface. The ionic strength of the liquid and the composition of the liquid must be considered in the decontamination process. We need to ensure that there will be ample contact time between microbes and bacteria so that there will be e better results. We can use electrostatic dispersion, immersion and spraying to enhance the efficiency of decontamination. Some of the decontamination systems affect the quality of meat products, and this will influence color, smell, and acceptability by the customers. It is worth noting that the decontamination systems must always reduce bacterial loads. We can use citric acid, acetic acid, and lactic acid at Limited concentrations so that the actual nature of the meat products is not disturbed. Ozone, hydrogen peroxide and inorganic phosphates can be used to decontaminate the carcass. Trisodium phosphate can be used to decontaminate poultry carcass effectively and there will be Limited rejection rates by consumers. We can use 10% of alkaline solution for this purpose also. There will be the formation of hydrogen peroxide radicals that can easily damage lipids, proteins, and nucleic acids. Hydrogen peroxide molecules exhibit both bacteriostatic effect and bactericidal effect (Lillard, 1994).

The chlorine dioxide and chlorine can be used in the decontamination process. We can use aqueous chlorine in the food processing industries to regulate the growth of microbes. When there is a decline in pH conditions, There will be a decrease in bactericidal activity. This can be attributed to enhanced organic matter proportions. There will be formation of reaction compounds that can be mutagenic in nature. Trihalomethanes can be mutagenic in nature. Chlorine dioxide has many advantages over aqueous chlorine. The action of Chlorine dioxide is not affected by altering alkaline conditions, and this is true in the case of organic matter. The use of antibiotics can also enhance the hygiene levels of meat products. We can use oxytetracycline and chlortetracycline also. We can use a combination of antibiotics to bring about good results. The use of antibiotics in preserving hygienic and healthy meat products has faced protest and criticism among most of the societies. The antibiotics that are given to decontaminate meat products can also bring about allergic and toxic reactions in the human body. In some cases, humans develop bacteria that can be resistant to antibiotics that are previously taken in the form of meat products. It is important to impose strict regulations on the usage of antibiotics in meat products. Stiffness of meat products results due to glycolysis that may occur about the death of meat-producing species and this can be called rigor mortis. This can be due to loss of ATP molecules in the anaerobic glycolysis process. Once ATP molecules are not available there will be rigor in the meat products. It should be noted that the meat products before rigor mortis can exhibit increased water holding capacity, and there will be emulsification of fat than meat products that have gone through rigor mortis. The pre rigor meat products can be easily processed (Valentina et al., 2010). This can be achieved at −20°C or below that temperature. The cold shortening method can be used in this case to minimize tenderness in meat. We can also use electrical stimulation to increase postmortem glycolysis. The electrical stimulation process can increase contraction relaxation cycles that can encourage postmortem glycolysis. We can also increase the quality of meat by using an electrical stimulation process. The palatability color and tenderness of the meat products can be enhanced using this technique. In the packaging sector, this process is especially important. In the meat industries, the exudative, soft and pale conditions are of primary concern; this is very important in the case of pork-based food industries. The combined preservation methods can be advantageous than other methods. These methods can be used to produce safe and stable foods. Some of the methods that can be used in combination are oxygen removal, smoking, fermentation, acidification, addition of Sugars, salting, curing, drying, freeze-drying, freezing, thrilling, and heating.

We need to follow good manufacturing practices in order to bring about safe and hygienic food for the customers. Good manufacturing practices can be named as current good manufacturing practices by many researchers. These are nothing but a series of Administrative procedures and manufacturing protocols that can regulate the products according to customer specifications. The good manufacturing practices can spread to several industries starting from veterinary products to pharmaceutical products.

Chapter

5

Biopreservation of Egg Products

CONTENTS

5.1. INTRODUCTION

The egg products can be considered as important nutritious food for most of the population. It is cheaper and many of the populations depend on it as food material on a daily basis. It is always desirable to increase the shelf life of the egg products through biopreservation (Figure 5.1) whenever possible. This is due to the fact that their products can be stored at the sight of customers for many days. In order to satisfy these requirements, the preserved egg products should not give any undesirable flavors in storage. The egg products can be stored and are treated with some of the specialized chemicals in order to extend the shelf life. Eggs can be stored at 4° centigrade to reduce the growth rate of microbes. We can store egg products at cool temperatures and then extend the shelf life for many weeks. This span of shelf life of any stored egg product can be determined by bacterial load and some of the organoleptic properties that we consider. We can treat egg products to eliminate microbes using different methods. Egg products can be heat treated or can be subjected to pasteurization. Harmful microbes like Listeria and Salmonella can be controlled by the pasteurization process. This process is safe, and there won't be any undesirable changes in the egg products.

Figure 5.1. Biopreservation of eggs.

Source: https://practicalselfreliance.com/ways-to-preserve-eggs/.

We can use electromagnetic radiation to eliminate several microbes and cause undesirable changes in the egg products. We can use Gamma radiation and radio waves to reduce the microbial growth rates. There are several antimicrobial agents that can prolong the shelf life of most of the egg products. These antimicrobial agents can function as static agents and cidal agents. Cidal agents can kill the microbes, whereas static agents can keep microbial growth at a constant rate (Antonio et al., 2014).

5.2. EGGS

A chicken egg can be classified as food with high nutritional value as it contains many nutrients, proteins, fatty acids, minerals, and vitamins. These nutrients are assumed as important perfect functioning of the human body. The nutrients provided by eggs can be easily observed by the human body. It is available in most of the countries at affordable cost. Consumption of egg has increased worldwide because of low cost and having high nutritional value. The consumption of eggs increases every year all over the world, and the demands for increased production rates. China, United States, India, Mexico, Brazil, and Japan are the largest producers of eggs. China produces at least 36% of total egg production. We can observe 12% ash, 11% lipid, 12% protein, and 65% water in eggs. In addition to the above components, there is Zn, copper, phosphorus, iron, calcium, and other minerals. We can also find Vitamin D, pantothenic acid, riboflavin, ascorbic acid, tocopherol, and retinol in eggs. We can also find several fat-soluble and water-soluble vitamins in this product. Some of the amino acids that are particularly important for human beings are present in eggs. Egg primarily constitutes three important components namely yolk, egg white, and shell. 31% of egg is made up of yolk, 58% egg white, and 11% shell. The egg white contains 8% of water and it has high proportions of protein and limited fat. The egg white also known as called albumen, has at least 40 different proteins. The antimicrobial properties and functional characteristics of the egg white can be attributed to these proteins. Important proteins that are present in egg albumin are ovoglobulin, avidin, ovomucin, lysozyme, ovomucoid, conalbumin, and other proteins. There will be presence of carbon dioxide in the egg white, which make them appear white at this substance is absent in older eggs. This is a reason why the stalked eggs appear transparent then the fresh ones. Albumen can easily obtain from the eggs, and this is the reason why it is being used in the preparation of several food items. The temperature for denaturation of egg proteins is around 58 degrees centigrade.

The yolk of an egg can be classified as a centralized position in the egg white. Generally, it is yellow in color and it accounts for 31% of the total heat egg content. Yolk of egg accounts for three quarters of the total energy of the egg. Pigmentation in the egg will be dependent on the feeding material in the poultry. The pigmentation difference does not affect the nutritional quality of the egg. Most of the nutrients of the egg will be concentrated in the yolk. It is mainly composed of lipids and proteins. Mostly we can observe globular proteins and lecithin in the egg. All the emulsification reactions that occur in the egg environment is dependent on lecithin protein (Petrak et al., 2000). The yolk contains 50% water, and the denaturation temperature is at 62.5-°C. Some of the processing operations are packaging, drying, freezing, cooling, pasteurization, stabilizing, blending, filtering and breaking. Eggs are being consumed Worldwide because of their versatility and its use in culinary preparations and services. Important characteristics like emulsification, gelatinization, and incorporation of air attract most of the food industries to choose egg related products. Eggs will be subjected to washing, packaging and also classification. Storage of eggs demands several challenges in order to maintain its quality. Due to the presence of rich nutrients in the egg, they are susceptible to several pathogenic microbes and spoilage microbes. Some of the eggs will not be subjected to the complete food processing steps as they are consumed naturally. It is essential to know that the quality control is not mandatory in this case. The quality of eggs will be visible only to the consumers who use it. Eggs should be stored at specified temperatures and for stipulated time so as to maintain quality and safety. Most of the eggs pop act in the natural forms as this is of primary concern for the food-based industries. There will be contamination mainly due to oviposition, and this must be carefully monitored by the poultry managers. Eggs come in contact with fecal material and hence the microbes contaminate them. Microbes mainly enter the eggs through microscopic cracks and pores. This contamination is possible during storage, packaging transport and washing process. Some of the microbes that can affect the egg are Listeria monocytogenes, Salmonella, and Staphylococcus species (Nemeth et al., 2011).

Figure 5.2. Pasteurization.

Source: https://www.zwirnerequipment.com/blog/methods-of-pasteurization/.

5.3. PASTEURIZATION

Pasteurization (Figure 5.2) is an important method that can be used in treating eggs. This will extend the shelf life of eggs. We can use a thermal pasteurization process to reduce the microbial load in the eggs. The egg that is pasteurized can be used in the food industry. This method saves, labor, and time and can be managed to have a limited space to store a large number of eggs. This method can reduce the functional properties, nutritional value, color, and flavor of the eggs. The freeze-drying method (Figure 5.3), radiofrequency, microwave, steam at the hot water can be used in decontamination of eggs. Each method can use diverse temperature conditions. The decontamination methods that need heat can be classified as efficient methods compared to methods that do not require heat. There are certain drawbacks with the methods that involve heat treatment. They can leave food less attractive to the end users because the texture and color of the food products will be damaged. These sensorial properties can be deeply affected due to usage of excess heat.

Figure 5.3. Freeze drying.

Source: https://acrelife.com/how-to-freeze-dry-eggs-at-home.

The hot water method is another important method that can be used to decontaminate eggs. Hot water can be used as a primary option in the pasteurization process. It is important to know that a limited proportion of egg ingredients can be pasteurized with this method. We can use hot water to reduce microbial load in eggs using a water bath. In this method, water can be heated to a specified temperature, and eggs will be added to the hot water with different time intervals. Due to the pasteurization process, using circulated water at 57°C can bring down Salmonella typhimurium species and the exposure time was standardized at 15 min in the water bath. At this temperature, the quality of the egg was not disturbed, but there was a drastic reduction in the microbial loads.

5.4. STEAM METHOD

The steam method can be used to reduce the microbial load of eggs. We call it the steam pasteurization process. This process can be assumed as an alternative for the surface decontamination process. Steam gun treatment has been proved effective to decontaminate eggs. We can use steam generators to pasteurize the whole egg. This treatment can effectively reduce Salmonella species. The microwave method can be linked with the thermal method to bring about effective decontamination of eggs. In this method, there will be e volumetric heating, and it will be at Rapid rates. The microwave frequency that is in between 300 GHz to 300 megahertz can be used for the decontamination process. Most microwaves can interact with direct materials and improve the random motion of molecules due to electromagnetic disturbances. Most of the foods represent the absorption capacity of microwaves, and this can be attributed to the presence of excess water molecules. In some cases, the presence of carbon molecules can also help in absorbing microwave radiation in food products. Due to these features' microwaves can be used to dig to decontaminate egg and other food materials. This method is economical and can be used easily. The observed energy can be distributed all of the food matrix, surface area of the food products. If high amounts of energy are applied, then the egg loses its integrity. It is always important to use low levels of microwave radiation in the decontamination process, especially eggs. The membrane of an eggshell can be transparent to microwave radiation. Radiofrequency heating can be used to minimize the microbial loads effectively. The radio frequency band can be in kilohertz instead of megahertz mostly. Radiofrequency heating can be called high frequency dielectric heating. The freeze-drying process is another important process that can be used for the same purpose. We can preserve eggs through the dehydration process. This process can drastically reduce the space required to store eggs for prolonged periods. There are several preservation techniques that can be used to preserve the quality of food without disturbing the shelf life.

5.5. DEHYDRATION

Dehydration can be an important process in preserving eggs, and it has several advantages. In this process, the microbial growth will be stabilized, and it will be easier to use because of its uniformity. The eggs can be easily transported, and it also requires Limited space. We can use cryodesiccation or freeze-drying to dehydrate the egg, and can convert egg products into powder. In this technique, we need to use temperatures that range between −50 to – 60°C. This process can be called the Rapid freezing technique. The water content that is present in the eggs will be subjected to this treatment, and hence it transforms into a solid-state forward by vapor state, and this process is assisted by sublimation technique. There will be specific vacuum conditions and low temperature conditions that are to be maintained. The lyophilizer or freeze dryer can be used in the freeze-drying process. In this process, we need to use a vacuum chamber, vacuum pump, condenser, and heat source. The important function of the vacuum chamber is to tolerate pressure variations so that there will not be any melting of ice. Some of the gases that are not condensable can be removed in this process. It is important to install a heat source that can produce energy and involves evaporation of ice. The type of source must be considered to define the freeze dryer type. The condenser can retain moisture and the food products. We can use the freeze-drying method in three different phases. At first, we need to initially freeze the act. There will be a primary drying process in which water will be eliminated. This step may need vacuum conditioning and some heat will be added. The next step will be secondary drying, and this can be called as desorption. This step may occur after eliminating all the available ice from the environment. The liquid water will remain in the egg itself; at least 2% of the moisture will be reduced. To reduce moisture, we need partially dried eggs that are placed in a freeze dryer for at least two hours. Due to the freeze-drying process, most of the nutritional quality and sensory characteristics will be retained as the temperature is not exceedingly high. The shelf life can be extended if the packaging is done perfectly without any flaws. Most of the volatile compounds will not be retained by water vapor and hence stay with the food matrix. This will ensure that the aroma of the egg will be retained. Due to this process, the protein structure will be modified. The tertiary structure and the quaternary structure of most of the proteins present in eggs will be altered. The proliferation of microbes will be altered because of reduced humidity. Most eggs will suffer from oxidation reactions that involve aromatic substances, vitamins that are fat-soluble, carotenoids, and lipids. It is to be noted that the freeze-drying process is costly and time-consuming process.

5.6. HHP

The High hydrostatic pressure technology (Figure 5.4) can also be used for the processing and preservation of eggs. It is also called HHP Technology. With this technology, we can damage microbes even at moderate temperatures. We can preserve bioactive nutrients and enhance bioactive compounds that are to be extracted. Most of the allergic substances that may arise from the food products can be regulated. In this technology, high pressures will be applied that range between 100 megapascals and 1000 megapascals. At this pressure, most of the microbes will be eliminated at which guarantees microbial safety standards for the end users. This technique can be used in the form of a batch system. We can use water as a transmission medium that can carry pressure onto the food products that are to be processed and then decontaminated. Most of these food products will be packaged and kept under heavy pressure (Nassim et al., 2017).

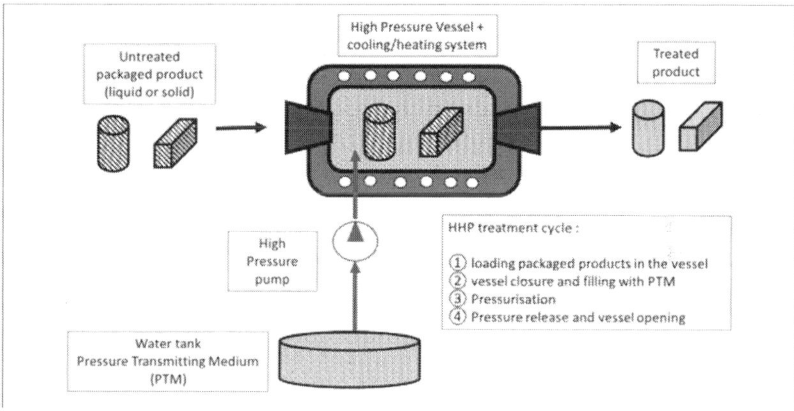

Figure 5.4. High hydrostatic pressure technology.

Source: https://www.researchgate.net/publication/330448697_Potentialities_and_Limits_of_Some_Non-thermal_Technologies_to_Improve_Sustainability_of_Food_Processing/figures?lo=1.

This technology mainly uses alloys made up of high strength steel and this equipment is resistant to rupture and oxidation. This equipment can be classified as batch equipment and in some cases, we can classify it as a semi-continuous system. This batch equipment constitutes a cylinder made up of thick walls and we call this cylinder a Pressure vessel. This batch system can be used for solid foods and liquid foods, and in most of the cases, we can use this for foods that can be pumped easily. In this process, food can be sealed and packed into small baskets. Most of this packing May contain flexible materials, and most of these materials are resistant to the pasteurization process. These small baskets can enter the pressure vessel and this visual has transmitting fluid that can hold pressure. We can use water to transmit pressure at an industrial scale. We can use an intensifier and pump to control pressure with the help of compression. It is worth noting that the product will be subjected to appropriate pressure for stipulated time. After the entire process is completed, we can depressurize the vessel. HHP technology can be used to bring about biological, chemical, and physical changes in the food products. Some of the changes that can be observed are altered enzymatic activities, protein denaturation, temperature, time of treatment and pressure. The biochemistry, cell membrane, and morphology of the microbes will be drastically altered by an application of HHP Technology. The cell membrane can be targeted in this technology. This technology can be used in most of the food processing industries as it destroys microbes and preserves sensorial and nutritional characteristics along with extended shelf life.

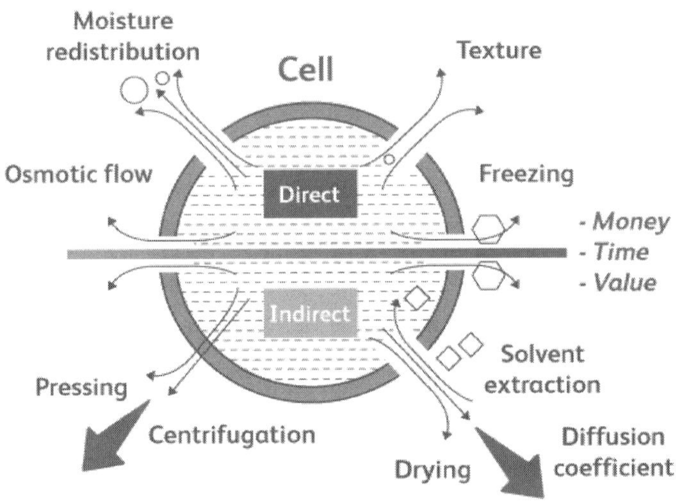

Figure 5.5. Pulsed electric field.

Source: https://www.pulsemaster.us/pef-pulsemaster/faq.

5.7. PULSED ELECTRIC FIELD

A pulsed electric field (Figure 5.5) can be called a high intensity electric field that can be used in the food processing sectors. We can use this method to preserve foods that are in the liquid stage. With this technology, we can preserve vitamins, texture, color, and flavor. Most of the enzymes produced by microbes can be inactivated, and there will be death of several microbes. Repeated electric fields will be supplied to food products within short time intervals so that most of the microbes will die and several enzymes will be inactivated. In this method, High Voltage pulses will be used along the two electrodes that were installed in the food products. The electric intensity will be higher, and there will be an increased amount of energy in the condenser. Mostly there will be discharge of energy in the form of pulses, and the supply will be in uniform nature. There will be a limited increase of temperature during this process.

Chapter 6

Biopreservation of Sea Foods

CONTENTS

6.1. INTRODUCTION

Biopreservation of seafoods (Figure 6.1) can be very important as most of them are susceptible to microbial contamination during several stages of transport and processing. The seafoods can be classified as food commodities that are fragile in nature. Some of the organoleptic, microbial, and nutritional qualities of seafood and fish best products can be easily spoilt due to microbial action. It is always necessary to monitor microbial quality in all the stages of food processing, and this is possible by hurdle Technology and that assumes safe transitions are food products from catch to plate. We can use mild preservation techniques like biopreservation to safeguard several food products that are classified as seafoods. The antimicrobial metabolites and natural flora along with lactic acid bacteria, must be studied to understand the biopreservation process (Peter, 1994).

Figure 6.1. Biopreservation of seafoods.

Source: https://www.agriculturenigeria.com/manuals/processing/livestock-processing/fish-preservation/.

Almost 35 types of lactic acid bacteria exhibit antimicrobial activity, and they can act as protective agents especially in the fishery industries. The lactic acid bacterial species must be selected best on specific criteria. This criterion includes resistance to antibiotics, production of biogenic amines like histamine, cross inhibition, super chilling process, resistance to coating using chitosan and antimicrobial activity. We can use Serratia proteamaculans (Figure 6.2), Hafnia alvei (Figure 6.3), Lactobacillus sakei (Figure 6.4), Brochothrix thermosphacta (Figure 6.5), Photobacterium phosphoreum (Figure 6.6) and Shewanella baltica (Figure 6.7) species can be inhibited if lactic acid bacteria were used.

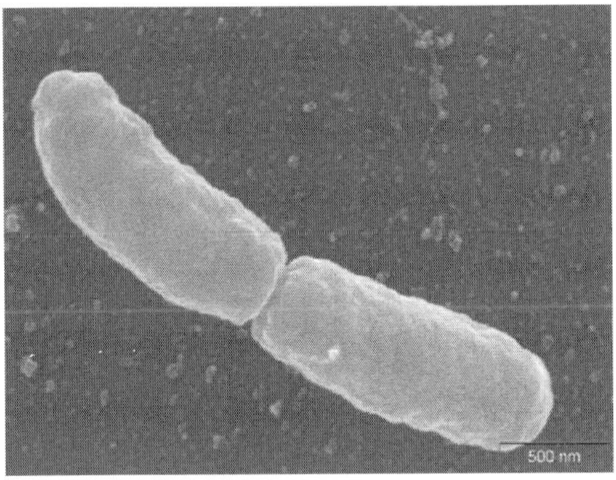

Figure 6.2. Serratia proteamaculans.

Source: https://genome.jgi.doe.gov/portal/sheb3/sheb3.home.html.

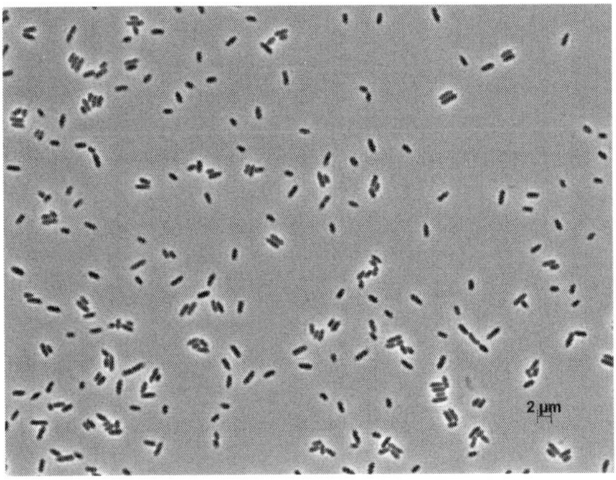

Figure 6.3. Hafnia alvei.

Source: https://bacdive.dsmz.de/strain/4938.

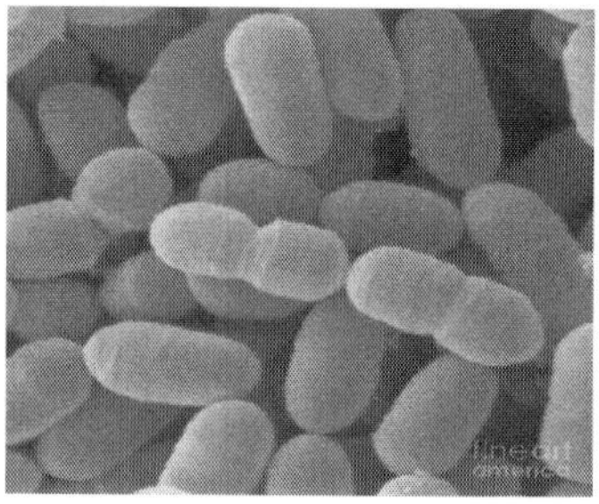

Figure 6.4. Lactobacillus sakei.

Source: https://fineartamerica.com/featured/3-sem-of-lactobacillus-sakei-sci-mat.html?product=poster.

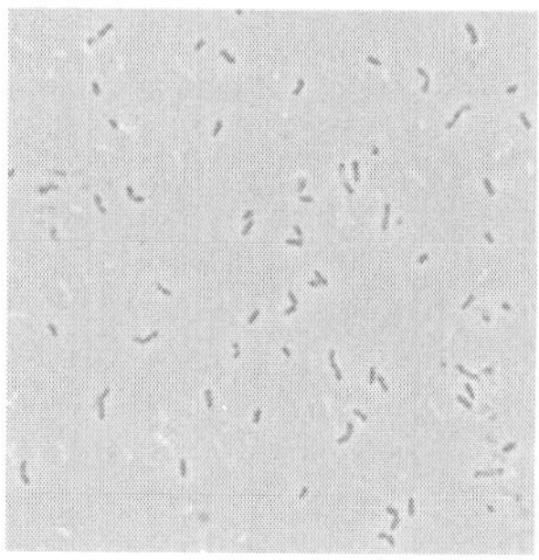

Figure 6.5. Brochothrix thermosphacta.

Source: https://atlas.sund.ku.dk/microatlas/food/bacteria/Brochothrix_thermo-sphacta/phasecon.html.

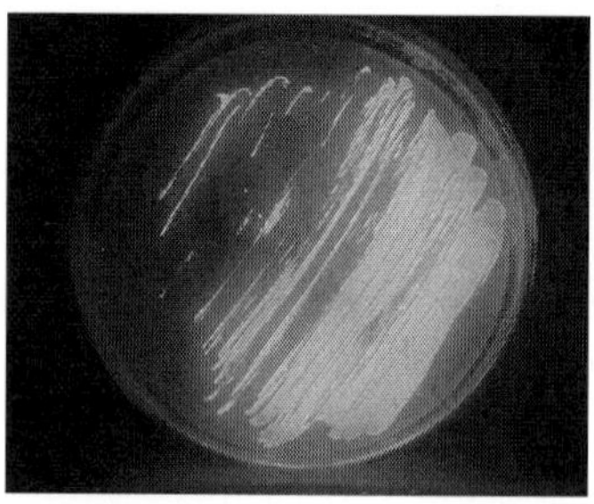

Figure 6.6. Photobacterium phosphoreum.

Source: https://www.agriculturenigeria.com/manuals/processing/livestock-pro cessing/fish-preservation/.

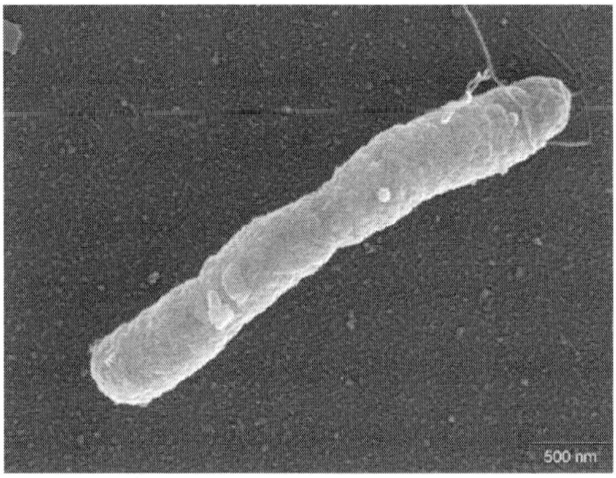

Figure 6.7. Shewanella baltica.

Source: https://genome.jgi.doe.gov/portal/sheb1/sheb1.home.html.

Seafoods contain several valuable nutrients like Omega 3 fatty acids, minerals, vitamins, and proteins. There was an exponential increase in India's aquaculture production and fisheries. Most of the fishery products can be brought under fragile commodities. They have a short life of 14 days. This can be due to an increase in pH values towards acidic nature. There will be nitrogen content with limited protein concentration. Trimethylamine oxide can be present in fish products. In the fish products, amino acids such as cysteine and methionine can be present, and they create undesirable flavors and orders also. This nutrient whose composition of flesh Can attract several microbes therefore degrades the sensorial characteristics of foods.

6.2. SPOILAGE MICROBES

Most of the specific spoilage microbes can get attracted to the nutritional quality of the fish products. These microbes can degrade the sensory attributes of the seafoods. The lightly preserved foods and freshly preserved foods can get spoiled due to psychrotrophic bacteria that is gram negative bacteria. Some of the bacteria that can spoil these products are Brochothrix thermosphacta, Hafnia and Serratia species, Photobacterium species, Pseudomonas, Aeromonas, and Shewanella species. Some of the lactic acid bacteria can also spoil the seafoods. Some of them are Lactobacillus plantarum, L. farciminis, L. sakei, and L. curvatus. Most of the seafood products can act as microbe carrying vectors. We can declare that at least 10% of the diseases associated with foodborne illness can be linked with seafoods.

Staphylococcus aureus, Salmonella, Vibrio and Listeria monocytogenes can spoil the seafoods and can be treated as a potential risk of humans (Birgit et al., 2006). The regulation of pathogens and food spoilage must be treated as an emergency research that is to be considered by most of the food agencies. The microbial risks can be reduced, and shelf life can be extended by the usage of active packaging, chitosan coatings, super chilling, pulsed electric fields, HHP, and biopreservation. We can consider biopreservation as the application of bacteria and its metabolites to reduce the microbial loads. The lactic acid bacteria can be present in most of the food products. They can produce reuterin, hydrogen peroxide, acetaldehyde, diacetyl, fatty acids, organic acids, and bacteriocins. Most of these compounds are non-toxic and can be consumed by humans with limited intervention.

Fish products can induce several beneficial effects on the heart tissues, and most of the chemicals produced from the seafoods can be anti-inflammatory in nature. Some of the useful compounds that can be generated from fish foods are eicosapentaenoic acid, docosahexaenoic acid, and omega-3-polyunsaturated fatty acid. It is important to eat fish products at least twice a week for good health. The fishes may not be stored in normal conditions. Due to endogenous enzymatic activity and attack by the microbes, fishes are more susceptible to damage. It is always important to decontaminate the fish products as soon as the harvest is made. While fish products are beginning to get spoiled, there will be several new compounds that might be detrimental for human health. It is always important to develop methods to increase the shelf-life period. There will be oxidation of lipids, and most of the proteins will get degraded during the decaying process. We need to keep in mind that the mushy and soft texture of the fish makes it vulnerable to several microbial attacks and hence easily spoiled. The moisture content, microbial proteases that are endogenous in nature, oxygen and holding temperatures can affect the decaying process. These observations can be made while postmortem handling. These factors can damage the important characteristics like flavor, texture, odor, and color. The fishes can be stored in the ice slurries, refrigerated seawater, in flake ice and in cooled cabinets. There are several natural substances that were developed using fungi and bacterial species. This development surfaces when the customers are worried about the role of chemical preservatives in the human body. Most of these chemical preservatives are harmful to humans. Most of the natural preservatives were used in the food-based sectors. Most of the natural preservatives can be made from fungi, algae, bacteria, animals, and plants. The chitosan molecules can be generated from the shrimp shells and crab. These can be classified as preservatives that are derived from animals. The rosemary extract, tea polyphenols, and thyme oils can be derived from plants. The bacteriocins can be derived from the microbes.

There are three major reasons that can describe the spoilage of fish products. They are microbial growth, oxidation, and enzymatic autolysis. Most of the degradation process in the fish products and freshly harvested fishes can be due to autolytic enzymatic spoilage. Most of the major molecules will cleave the macromolecules. These enzymes can degrade the texture of the fish products, and this can be called a softening process. The softening mechanism can damage the whole texture of the fish in minimal time. The off-odors and off-flavors may not be produced due to autolytic enzymatic breakdown. The quality of the fishes will be reduced even with limited enzymatic activity. This will also lower the actual shelf life of the fishes. In the fresh harvest of the fishes, we can observe endogenous muscular enzymes and digestion enzymes of the gastrointestinal tracts. These enzymes can be observed in the muscle tissues also. These enzymes were studied in detail and conclusions were made to explain that these enzymes have an immediate role in the postmortem damage of the fish products. The activity of these enzymes is evident in the processing steps and while the storage phases. The sensory characteristics of the fish products will be affected due to the action of these enzymes. There will be autolysis of the muscle proteins of the fishes, and this can result in the production of the biogenic amines that may spoil the fish products. It is always important to define the stage of autolysis prior to the onset of the spoilage process. The fish products should be subjected to reservation at low temperatures to ensure that the enzyme activity is kept at a minimum. The enzymatic activity at the local tissues can be checked due to exceedingly low temperatures.

6.3. LIPID OXIDATION

The oxidation of lipids can also deteriorate and spoil the fishes. The peroxidase activity, photosensitized oxidation reactions, microsomal enzymes, lipoxygenase activity and auto-oxidation mechanisms can spoil the fish products. The lipid oxidation can be initiated by the release of transferrin, hemosiderin, myoglobin, ferritin, and hemoglobin while in the storage phase. We can classify the oxidation of lipids into enzymatic and nonenzymatic reactions. These reactions can be detrimental to the fish quality. The fatty acids will be subjected to oxidation, and this will cause severe problems with the fish harvest. Mostly, the double bonds present in the fatty acids will be targeted by the oxygen molecules. It is obvious that most of the fishes contain unsaturated fatty acids. These products can also exhibit high levels of prooxidant molecules. Due to the presence of these molecules, these fatty acids can be easily exposed to the oxidation process. This can lead to rapid oxidation, and hence there will be loss of quality and rancidity may develop easily. The whole texture of these products will be disturbed. The nutritional values, texture, color, and smell will be disturbed. The oxidation of lipids can result in the development of fluorescent compounds, loss of antioxidant systems that are located endogenously, loss of nutrients and protein denaturation. Mostly, we can conclude that fish loses its native property due to the action of these molecules during the storage phases.

It is common opinion in the public that fish meat helps in maintaining human health well inside the medical limits (Haultin, 1994). The fish products are popular because of their limited cholesterol components. We can observe high levels of protein components and amino acids that can help humans in various ways. Some of the important amino acids that can be present in the fish products are omega-g and omega-3 fatty acids that have important medicinal values that we can reap. The vitamins, minerals and fatty acids can be present in the fish meat. It is worth noting that the fish products serve as an important dietary staple in most of the countries harboring waters in the form of rivers or seas. The tuna products were studied to establish an advanced processing setup to preserve food products per se. Most of these advances are limited to the canning of these fishes. Due to overwhelming demand from the fish-eating communities, there was a need to develop the latest food products that can satisfy the customer's taste. The lipid and protein components in the fishes can define the sensory and nutritional properties of the fishes. Due to antimicrobial and antioxidant properties of clove, we can use it in the preservation of the fishes and these properties can be attributed to the phenolic compounds present in cloves. The viral contamination that can occur through food can be minimized with the use of ginger and clove extracts.

These extracts have attractive chemical profiles that make most of the food researchers study them in detail. Kiwi fruits are also being used for preservation of fish products and these fruits are popular for their nutritive and sensory attributes. The garlic and ginger extracts can exhibit firm antimicrobial and antioxidant activity, and these extracts can be used as natural preservatives. This kiwi fruit is found in almost all areas though it originated from the Asian countries.

We can obtain minerals, vitamins, fat, and proteins from echinoderms, molluscs, crustaceans, and fishes. They are of high nutrient value and they serve as delicacies. The shelf-life of these seafoods is limited, and this can be attributed to the increased amounts of moisture, neutral pH, and other essential minerals. The nutritional and sensory properties of these seafoods can be disrupted after the death of the species. The lipid oxidation reaction is high in the fish products because of the presence of polyunsaturated fatty acids. The color changes in the seafoods can be attributed to the generation of unhealthy molecules that can result in nutrition loss, loss of flavor and odor. These seafoods will become unpalatable due to the production of unhealthy compounds. The spoilage of seafoods can be physical, chemical, and biological, and this leaves the food spoilage a complex process to understand. The chemical reactions and indigenous enzymes can act on the fish products and they get spoiled easily. Most of the spoilage can be attributed to the presence of microbes for prolonged times. The distance from the area where harvest is done and the processing stations can be classified as an important factor that defines the quality of the seafood. Along with this, we need to consider storage temperature and processing facilities into consideration. The intrinsic and extrinsic factors and the methods that were employed to catch fish and other marine life can be important too. These methods can both positively and negatively affect the food products. All these factors can affect the shelf life of the fish products. The good management practices, appropriate hygiene, and critical control points can be moderated according to the fish type and their native conditions. The shelf-life of the food products can be extended, and there was a grown interest in this area. Most of the population are demanding for the fish products due to their health benefits. These products have become a popular brand among several business owners who focus on food-based industries. We can use sulfur dioxide; sodium nitrite and sodium benzoates as chemical preservatives as natural preservatives are sparingly available. The main problem with these chemical preservatives is that they can be detrimental to human health. We can use salt to preserve fish products, and it can be classified as a conventional technique that can work better than its counterparts. Usage of salt can be an economical process and may not attract more microbes. This technique is simple and can be easily executed in a common processing unit with no hurdles whatsoever. The water activity can be effectively reduced by the addition of salt. This is true in the case of seafood muscles of the seafood. The enzymatic activity and bacterial growth will be hindered. Some bacteria that were classified as halophilic bacteria and halotolerant bacteria may easily grow in the seafood that were preserved using salt. These microbes can use energy to avoid the salt content in the cells, and thereby, the aggregation of the protein content can be halted. This process can be called the salting-out process. This mainly involves cytoplasm, and mostly this salting-out process can be observed in microbes that attack several fish products. Due to this process, we can observe enhanced sodium concentrations in the fish products. If these microbes are left without any treatment, they can spoil the fish products that are stored, and this can occur due to the fermentation process. The taste of the fish products can be hindered due to the salting-out process and fermentation thereof. We can use bioactive peptides, chitosan, bacteriocins, essential oils, and plant extracts as preservatives, and they can be classified as natural preservatives. These natural preservatives can easily replace the synthetic preservatives as they can exhibit antioxidant and antimicrobial activities. These natural preservatives can also withhold the fish from being deteriorated by the microbial action. Most of the lipid oxidation will be ceased due to the addition of these natural preservatives. The consumers and processors can remain profited due to the addition of these natural preservatives. They are safer to use, and they can also improve the quality of the seafood.

We generally refer to spoilage as a deviation from the original texture, appearance, taste, and odor that can withdraw the palatability of the foodstuff. There will be several attacks by the pathogenic microbes and other opportunistic microbes on the fish products. The habitat of most of the fishes can be opened to the growth of several microbes that we cannot control. The production of ketones, aldehydes, organic acids, sulfides, putrescine, histamine, alcohols, and other molecules in the fish meat by the microbes can be the cause of concern among several fish-based entrepreneurs.

We can observe an increased load of psychrophilic bacteria in the seafood that are refrigerated and chilled for storage. Some of the psychrotrophic bacteria can be aerobic in nature, and some of these microbes may exist as facultative anaerobes. These bacterial species can be gram negative in nature. Some of the microbes that can spoil the fish products are Vibrio, Flavobacterium, Aeromonas, Photobacterium, Pseudomonas, Acinetobacter, Shewanella putrefaciens and Moraxella. We can declare that most of the bacterial species that can damage the fish products are gram negative bacteria. It should be understood that the extended transportation times and processing periods can attract several bacterial species and hence the spoilage. P. phosphoreum and lactic acid bacteria can be assumed as major spoiling bacteria found in most of the fish products. Seafood can get spoiled by the action of Brochothrix thermosphacta, Bacillus and other species. We can assume that both gram-ve and gram +ve bacteria can spoil the fish-based products easily. The fishing methods, location at which the collection is made, and the location of sampling can define the type of microbes that infest the fresh fish products or stored products. The amino acids that are freely available in the fish carcasses, carbohydrates, and small peptides are of low molecular weight and they can be easily absorbed by the bacterial species. Some of the by-products that can be produced by these bacterial species are S compounds, histamines, and biogenic amines.

The trimethylamine oxide can be reduced by S. putrefaciens, Aeromonas spp, Vibrio spp and Enterobacter species into trimethylamine. This molecule is dominant in seafoods. This is the main reason why fish products have a characteristic odor. The production of trimethylamine can yield a bitter taste to the fish meat and this can worsen if there is generation of hypoxanthine. The hypoxanthine molecules can be produced by the activity of the indigenous enzymes, and their release is too quick. They can be produced by the action of bacterial species, and this can be due to degradation of nucleotides like inosine monophosphate and inosine. As mentioned elsewhere, the seafoods are highly packed with lipids, and most of them are polyunsaturated fatty acids with long chains. These lipids can also act as a mediator in withstanding the nutritional quality of the fishes and odor. The oxidation of the lipids can also result in the reduction or removal of the vitamins that are mostly fat-soluble. The oxidation of lipids can occur in many stages. The hydrogen atom that is labile will be present in the chain of the fatty acids, and it is present at the acyl chain. This will start the lipid oxidation reactions. In this process, there will be formation of free radicals. These radicals can readily react with the oxygen. This reaction will yield peroxyl radicals. These peroxyl radicals can attract H atoms that can be present in alternate fatty acyl chains. This will become a chain reaction in no time, and there will be the formation of several free radicals and many molecules of hydrogen peroxide. Once there are sufficient free radicals in the medium, the oxidation of lipids reaction will cease. Then, we can observe the generation of products that are nonradical in nature. The oxidation rates can be regulated by the degree of unsaturation, temperature, moisture, metal concentration, light, and availability of oxygen. In this process, the hydroperoxides that were formed can be assumed as unstable molecules. There will be formation of secondary products that can be generated due to degradation of the products that were classified as primary products. We can assume that both primary and secondary products will be produced in the lipid oxidation process.

Some of the primary products are peroxides, dienes, and free fatty acids. The secondary products can be carbonyls, trienes and aldehydes molecules. All these products can carry a certain grade of prominence in the spoilage processes. The magnitude of the production of these oxidation products and types can be based on the composition of the fatty acids and redox reactions thereof. The magnitude of oxidation and intensity can be dependent on the packaging methods, additives, reduction of sizes, temperature, processing parameters, post-slaughter activities and preslaughter activities. Some of the preslaughter activities that are to be considered are stress and physical injuries. The main post-slaughter activities that can be studied are pH, temperature, tenderization techniques and cold shortening.

There are several prooxidants that can initiate the oxidation of lipids. Some of the prooxidants that can participate in this reaction are Cytochrome c, myoglobin and hemoglobin molecules. The oxidized hemoglobin and deoxygenated hemoglobin can be present in the fish blood and it can enhance the rate at which the lipid oxidation reaction takes place. Most of the bleeding events can diminish the lipid oxidation process. This can be attributed to the lowered concentration of the prooxidants molecules in the blood volumes.

If the secondary oxidation products get reacted with the peptides, amines, and proteins, there will be loss of functionality, and flavors will be disrupted. The hydrolysis of lipids can yield free fatty acids and proteins. These molecules can get interacted with each other. These interactions can denature the sarcoplasmic proteins and myofibrillar proteins. Some of the enzymes that are present in the fishes are subjected to enzyme deterioration. The indigenous enzymes will degrade the proteins, and this process can be classified as autolysis.

Due to the onset of rigor mortis, the fish flesh can be opened for microbial infection and growth. The lipases and proteases can damage the quality of the fish products. These enzymes can effectively change the sensory characteristics of the fish products.

This occurs even in the frozen states and refrigeration stages but at slow paces. Due to the action of the microbial proteases and indigenous proteases, there will be rapid degradation of the proteins, and this can be high in the irregular storage conditions. We can observe phospholipase, lipase, trypsin, cathepsins and chymotrypsin molecules in the pyloric ceca, spleen and hepatopancreas of the seafood. Some of the pepsin molecules can be concentrated in the stomach areas of the seafood. We can observe belly bursts in fishes, and this can be due to enzymes that are active in their guts, and this will ensure that there is immediate decomposition of the proteins. Whilst processing and storage, the generation of formaldehyde molecules can bring about changes that induce toughening of meat, and there will be a change in the texture of the fish meat. All these activities and reactions will be based on the enzymes. There will be the formation of formaldehyde and dimethylamine due to demethylation of the TMAO molecules. Some of the formaldehyde molecules may exhibit cross-links with the proteins. This process can be mediated by methylene bridging. This can result in the toughening of the fish muscles. This can be attributed to the lowered water holding capacity. We can also observe melanosis in case of shrimps. This condition can be due to the activity of polyphenol oxidase and tyrosinase that are dominant in the shrimps. Most of the microbial growth can be due to proteolysis of fish products that generate peptides and free amino acids. Due to these reactions, there will be the production of biogenic amines. The activity of the protease enzyme can be affected by the pH and temperature. Most of the proteases can act normal at neutral pH conditions and alkaline conditions. TMAO can get transformed into TMA through a reduction process.

There will be production of other volatile molecules also along with the TMA molecules. This conversion and formation of accessory compounds can be attributed to the enzymatic activities. The enzymes can be microbial enzymes or microbe derived enzymes. Some of the volatile compounds can be produced by the metabolic reactions that might occur in the flesh of the seafoods and microbes. Due to the action of these enzymes in succession, we can observe increased pH levels. These abnormal pH values can ring about sensitive conditions in the seafood that might attract several microbes. Most of the fats present in the seafoods can be hydrolyzed by the enzymes like phospholipase and lipase. Triacylglycerol acyl hydrolase can activate the hydrolysis of triglycerides in the fish products. Some of the free fatty acids will be released in the seafoods, and there will be an oxidation process, and this will effectively lead to fishy odor. Mostly fish odor will be based on the polyunsaturated fatty acids and aldehydes present in the fish products. The intensity of the fish odor can be based on the bleeding time of the fish after harvest. The odor will be less in the fish that is enabled even after harvesting.

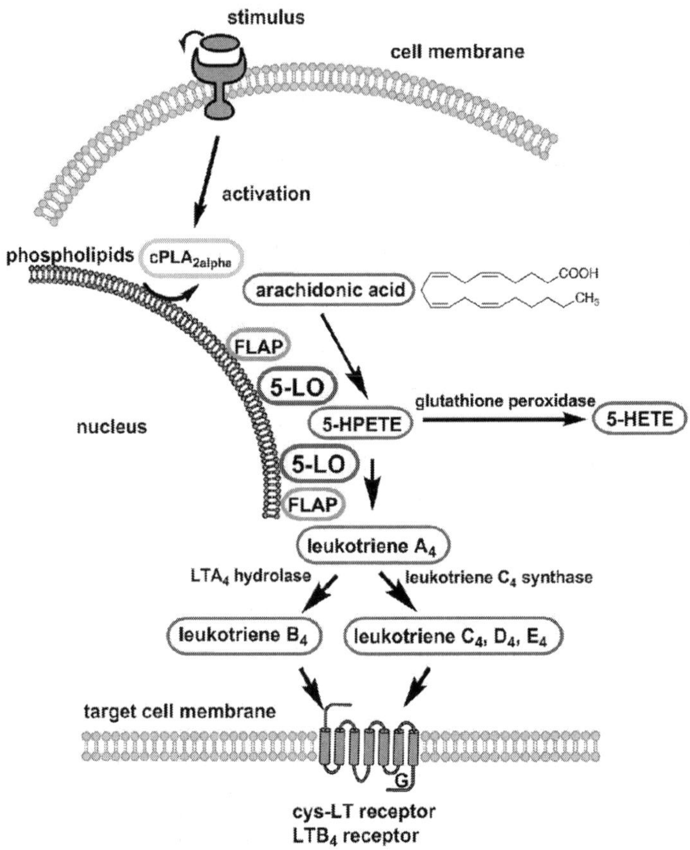

Figure 6.8. Lipoxygenase enzyme.

Source: https://www.frontiersin.org/files/Articles/8074/fphar-01-00143-HTML/image_m/fphar-01-00143-g001.jpg.

The lipoxygenase enzyme (Figure 6.8) will be dominant in the skin and gills and it is involved in the oxidation processes. The action of this enzyme will be enhanced if the storage time is extended. There was always a search for antimicrobials and natural oxidants in the fish-based industries all over the world despite their limited availability. Some plant-based extracts can be used to preserve fish foods, and they can be derived from peel, fruits, shrubs, and leaves. These extracts are dominated with the polyphenols. They can hold several phenolic rings that are active. The phenolic compounds were classified according to their activity and chemical composition. Broadly we can classify these compounds into flavonoid and nonflavonoid compounds. Based on the degree of oxidation, we can classify the flavonoids and anthocyanins, flavans, flavones, isoflavones, flavanols, etc. Mostly the degree of oxidation will be studied in relation to the heterocyclic rings. The nonflavonoids and phenolic acids are dominant in most of the plants though all plant extracts may not be used for preservation processes in the fish-based industries.

The antioxidant and antimicrobial activity of these plant-based extracts can be attributed to the presence of polyphenols, flavonoids, terpenoids, phenolic acids, ascorbic acids, tocopherols, retinoids, and carotenoids. Some of the essential oils exhibit similar properties due to the presence of the chemical components same as the plant extracts. The essential oils and plant extracts can contain aliphatic and aromatic compounds. These compounds can be derived from xylems, roots, stems, peels, fruits, leaves, and petals. The quality of the essential oils and their quantitative attitude can be based on the extraction process, extraction media, plant's age, climate, genotype, type of soil, provenance, and plant nature thereof. We can extract these compounds from the plants in two stages. The first stage can be called the preparation stage and the second stage can be called the extraction stage.

Some of the solvents that are generated from these extracts can remove chlorophyll from the leaves, and this will reflect an increased level of antioxidation property. Acetone can remove chlorophyll from the leaves, but the efficiency will be less when we compare it with the extracts and solvents thereof. The phenolic compounds can be soluble and can be separated from the residues. This can be done by the solvents that are specific for certain fish products. These processes can be done through standard processes. These processes will be observed in the extraction process. Some of the important extraction solvents are toluene, dichloromethane, hexane, isopropanol, acetone, ethanol, methanol, and water. We can also use microwave-mediated extraction, fluid extraction at supercritical conditions, accelerated solvent extraction, sonification, etc. Most of the phenolic compounds derived from the essential oils and plant extracts can disturb the integrity of the cell membranes. This can be due to interaction between the membrane proteins of fungi and bacterial species. Due to these interactions, there will be enhanced cell membrane porosity or permeability. This will lead to leakage of the K ions. Some of the cytoplasmic structures will be leaked, and hence there will be death. Most of the cell membranes of bacteria will be dominated by phospholipids. The compounds present in the essential oils and plant extracts will be hydrophobic in nature and they create a partition in between the phospholipids. This is the reason why the cell membranes will become weakened by the addition of these compounds.

The critical ions that are present in the bacterial cell will be leaked, and this will lead to the death of the microbe. It is to be noted that the cell wall of gram-positive bacteria is thicker than the cell wall of gram-negative bacteria, and this explains why the latter are more sensitive to these extracts and compounds. The extracts and polyphenols can also lower the oxidation reaction of lipids in the seafood. The antioxidant property of these compounds can halt the lipid oxidation mechanisms by interrupting the free radical formation. These antioxidants can consume species that otherwise can be involved in the peroxidation process. These molecules can be involved in the chelating process that contain metal ions, and hence the species that are reactive cannot be initiated. There will be scavenging of the oxygen molecules, and hence the production of peroxides will be slim and can be useful in withholding the flavor of the fish products. These compounds can easily cleave and disrupt the auto-oxidative chain related reactions. Most of the concentration of oxygen molecules will be reduced. The effectiveness of any antioxidant can be based on the disrupting range of the free-radical reactions. The H donors are mainly phenolic OH groups. Polyphenol can get converted into radical. There will be production of quinone formations. This will also delocalize the electrons inside the aromatic ring. These electrons will effectively stabilize the radicals that were considered as intermediates. Most of the polyphenols produced from the plants can stop or inhibit the lipid oxidation process.

Most of the naturally active compounds can be insoluble or sparingly soluble in water. These compounds can be less stable. The active molecules can get easily distributed onto the cells and target sites. The active agents can be entrapped using the encapsulation process. This occurs in the career matrix. This can also enhance the bioavailability rates. The stability differential of these compounds can be observed. There are many benefits of this encapsulation process. This will enhance the half-life of the molecules. The handling ability and bioavailability will be increased. The transfer rate of the active compounds can be reduced. The loss of these compounds due to evaporation can be minimized. The reactivity of these compounds can be reduced in the environment. The natural compounds can be encapsulated, and this will help in their distribution all over the food materials. We can use these compounds in limited concentrations and hence most of the unpleasant taste that can spoil the palatability will be minimized. When these compounds are encapsulated in the liposomes, the antimicrobial, antioxidant properties will be stabilized. This stability can be observed in case of Thymus spp. We can integrate plant extracts and essential oils into the food products. This can be done via films followed by wrapping. This can be called active packaging. In this process, these natural compounds can be released into the food products and can easily perform their function. The microbiological quality, physical and biochemical factors can also affect the natural compounds. Some of the physical factors that are to be considered are loss of weight and pH. Limonene can be used as essential oil that can be used in the preservation of seafood like Gilthead sea bream. On treatment with the limonene, the shelf life can be extended.

We can use Allyl isothiocyanate as a preservative to store Sparus aurata species. Allyl isothiocyanate can prolong the shelf-life of these products. The oregano oil can be used to store the rainbow trout species. The oxidation of the lipids can be analyzed using Thiobarbituric acid. This can be used as a reactive substance. When treated with oregano oil, we can reduce the load of Enterobacteriaceae members. The shelf-life of the food products can be increased to 21 days. We can use both these essential oils in combination to preserve the food products. The samples that are treated with these essential oils can reduce the load of the spoilage specific organisms. The silver catfish can be preserved using Aloysia triphylla (Figure 6.9) and this can be classified as an essential oil (Guerino et al., 2018).

Figure 6.9. Aloysia triphylla.

Source: https://www.thegardener.co.za/the-gardener/aa/aloysia-triphylla/.

This oil can be used in delaying the microbial infestation in the fish products. The filets that are prepared using Aloysia triphylla can exhibit prolonged sensory shelf life when treated with this essential oil. On application of essential oil produced from oregano, there will be limited growth of Salmonella typhimurium and Listeria monocytogenes. The shrimps that are coated with gelatin can be preserved with the solutions like essential oils from leaf and this has drastically reduced the melanosis process. The shelf life is increased, and most of the microbiological qualities are reduced.

There are certain limitations with the use of essential oils and plant extracts. The essential oils and plant extracts even used in minimal concentrations can induce antioxidant and antimicrobial activities at least in the in-vitro conditions. Rosemary can be used for this purpose. Plant extracts and essential oils might be needed in heavy concentrations than the chemical preservatives. The increased use of these natural preservatives may alter the flavors of the food products, and it is not acceptable. Organoleptically, this condition may not be acceptable by the consumers. The essential oils that can be added in minimal concentrations can be obtained from spices and herbs. They can contain a strong aroma, and this will affect the taste and aroma of the seafood.

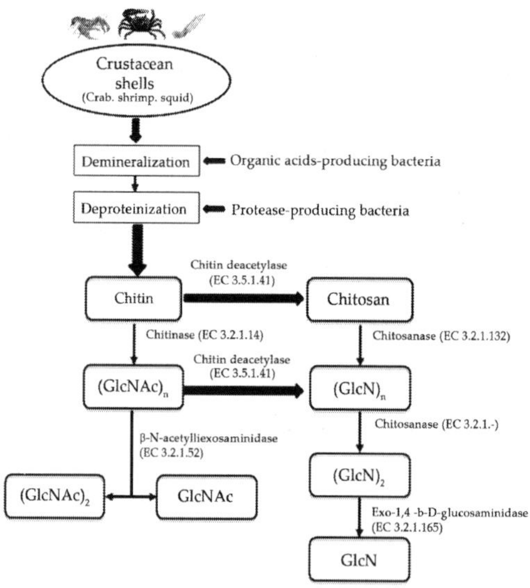

Figure 6.10. Chitooligosaccharides.

Source: https://www.mdpi.com/1660-3397/12/11/5328/htm#.

Chitooligosaccharides (Figure 6.10) and chitosan molecules can be used as preservatives. Chitin can be classified as an important polymer that is naturally available and it can be obtained from the shells obtained from the crustaceans. The chitin molecules can be subjected to deacetylation and there will be production of chitosan. Some of the chitosan molecules can be depolymerized. This can produce Chitooligosaccharides. These molecules can be called chitooligomers or chitosan oligomers. These molecules can be used in the fish-based industries due to their natural attributes, degradable and nontoxic nature. In these two molecules, there will be the presence of essential functional groups like OH group, acetylated amine groups and amines.

These molecules can easily interact with the cell receptors, and this will generate a cascade of reactions that can alter the organisms. Due to the presence of these functional groups, these compounds can exhibit antimicrobial, antioxidant properties and they are responsible for antidiabetic, immunostimulatory, neuroprotective, antiangiogenic, and anticancer effects. These compounds can get readily dissolved in the acidic solutions. The presence of water can further extend their applicability. These compounds can be used as film, coating agents and additive agents. These formulations can be used to enhance the shelf-life of the seafood. The chitosan molecules can be used in wrapping and coating.

Alkaline deacetylation can be used to produce chitosan from chitin molecules. This process may need 100°C temperature and requires alkaline conditions. This process might take several hours. The chitin deacetylases can be used for this purpose. There are diverse orientations of the microfibrils. Some of the polymorphic forms that can be observed are alpha, beta and gamma chitins. The alpha chitin molecules can be classified as antiparallel molecules. These are considered as stable forms. These chitin molecules can be present in the crustacean shells and chitinous cuticles. There are two primary methods phases that can be used in the chitin production. We can use the demineralization process to produce chitin from the shrimp heads. There will be a deproteinization process during the production of chitin molecules. There will be limited N-deacetylation in the production of chitosan molecules. The chitosan can be referred to as N-deacetylated derivative. The degree of polymerization can be less than 20 in the case of Chitooligosaccharides. In the case of depolymerization of these chitosan molecules, we can expect greater molecular weights accounting for 3900 daltons.

We can use physical methods and chemical methods that involve acids and enzymatic degradation. We can use nonspecific enzymes along with chitinases in this process. Some of the important enzymes that can be used in this case are lipases, proteases and cellulases. Mostly these enzymes can be used in the depolymerization process. We can also use the acid hydrolysis process for this purpose. Some of the chemicals that can be used in this case are hydrochloric acid, hydrofluoric acid, nitrous acid, and phosphoric acid. We can use ultrasounds that have low frequency, and this can be around 20 kHz. We can classify this technique as a physical method. We can use these compounds in the form of additives and coating agents. We can use these chemicals in the form of wrapping films. Most of these chemicals can be classified as environmentally friendly except a few. The chitosan molecules can be used as antioxidants, and they can work as oxygen barriers. The rate of respiration will be lowered with the application of these compounds. Due to these reactions, most of the microbes may perish without any further reactions and damage. The Chitooligosaccharides and chitosan molecules can exhibit several antimicrobial activities. Mostly they can be active against molds, yeasts and bacteria. The biopolymer can be positively charged, and the microbial cell membrane will be negatively charged and hence there will be interaction that can be mediated by chitosan molecules. There will be diffused hydrolysis that can interrupt the translation and mRNA will be inhibited thereof. There will be chelation of the spore elements, metals, and essential nutrients. The chitosans can exhibit a firmer reaction on the bacterial species than yeast and other fungal members. The bacterial species that were classified as gram-negative can be more sensitive to the action of these compounds. These compounds can act as primary and secondary antioxidants. There will be interruption of the radicals that can be called as intracellular radicals. The chitosan solution can be used as coatings to safeguard the seafoods, and this can be against microbial invasions and environmental factors.

Bacteriocins can be applied in major quantities in the fermentation process to regulate the microbial growth. We can use yeast extract and meat extracts as suitable substrates in this process. Biopeptones can be employed in the mass production of these bacteriocins. These substrates are costly, and they may not have wider acceptance because of the cost factor. We can produce bacteriocins by using fish meal, grape waste, and potato liquor along with other industrial wastes. We can also use molasses in this process. These compounds will have minimal environmental impacts and can be used easily. They are cheaper to use, and most of these compounds can be environmentally friendly in nature. Bacteriocins can exhibit excellent antimicrobial action against several bacterial species. We can use bioactive peptides for this purpose also. These are nothing but a minor fragment of the proteins that are being used in this process.

6.4. PEPTIDES

These peptides can act as immunomodulators and antioxidants. These peptides can be produced by the action of several proteases. We can use marine organisms, cereals, legumes, milk, egg, and fishes in the production of bioactive peptides. The fish processing by-products and wasted resources can be used as raw materials in the production of these peptides. Mostly it is preferable to use enzyme mediated processes in the production of these peptides. The specificity and proteinaceous materials can be classified as crucial factors in determining the bioactivities and amino acid sequences of these bioactive peptides that are being generated. We can opt for both alkaline and acid hydrolysis to produce peptides. These peptides may not be used in the case of food products due to certain limitations. The lysine amino acid can be converted into lysin alanine and tryptophan loss. We can also observe the racemization process. In most of the cases, the enzyme mediated hydrolysis will be because of proteases like ficin, pronase, alcalase, thermolysin, flavorzyme, neutrase and papain. These enzymes can be effective tools to produce peptides that can be completely functional. This is better than the hydrolysis caused by acid or base. The degradation of the proteins can result in the formation of the peptides, and hence their solubility will be increased. There are several peptides that are produced from the seabass and fish proteins. The concentration of enzymes, temperature, and pH can be important factors that are to be considered to understand the rate of hydrolysis. The degree of hydrolysis can define the quantum of peptides that are being produced.

The peptides can exhibit increased antioxidant activities, and there will be enhanced chelating potential. All these halts the oxidation of lipids. The copper (II) and Ferrous ions can serve as excellent chelating agents. This is the reason why they are being used in the food preservation process. These peptides can stop the formation of the free radicals and hence most of the reactive oxygen species will be stopped. This is the main reason behind using these peptides as antioxidants. Most of the free radicals that are formed through this process can react with almost all the amino acids. The sulfur-containing amino acids like cys and Met and aromatic amino acids like Trp,

Phe, and Tyr can exhibit maximum interaction with these compounds. There are certain amino acids with imidazole rings like His that can get reacted with the peptides easily. Most of the amino acids that are free in the solution and in the tissues might not act as antioxidants.

The fishpaste-based products can be classified as fish cakes and they can be stored as frozen products. These foods are popular in Japan and Korea. There are several food additives that can be added to store the food materials. Some of them can be produced from animals like chicken, swine, and beef. We can also produce these compounds from the seafoods like invertebrates and fishes. Cereals and legumes can be used in the production of these compounds that can be used in the preservation process. Polyols and sugars can be used in the storage of the fish products. There are several proteases that can be used as inhibitors, and they can be extracted from the animal sources. We can collect some of the plasma proteins from chicken, cow, and pig by-products. These proteins can be easily collectable and can be refined. We can extract beef plasma hydrolysate, bovine plasma protein, porcine plasma protein, chicken plasma protein, proteinase inhibitor, ovomucoid, egg white and egg albumin. These compounds can be used as additives. We can use the protein isolates that were extracted from the legume seeds in the form of protein additives, and this will effectively enhance the quality of the gels. Most of the isolates that were extracted from the legume seeds can contain trypsin inhibitors, and these compounds can be used as protein inhibitors. Some of these inhibitors can be extracted from the black cowpea seeds and soybean seeds. The protein isolates from the plants can be mainly used in the surimi-based industries as they are safe and cheap. Vicilin and legumin are the seed storage proteins that can be used as cogeling agents and binders. These protease inhibitors can act as binders and fillers.

The starch compounds can be used in fish paste-based products, and these products will be used to increase the weight of the product and its elasticity. The wetness, stickiness, and stability can be regulated based on the storage conditions. The surimi seafood can be modified by the addition of starch compounds as these compounds can alter the functional characteristics of these seafoods.

Chapter 7

Biopreservation of Fruits

CONTENTS

7.1. INTRODUCTION

There is always demand for the safe and nutritious fruits all over the world, and this needs certain synthetic preservatives so that they can be preserved for prolonged periods. But some of these synthetic preservatives can lead to palpitations, cancer, and headaches. The fruits can be classified as an important diet component in several communities. Watermelon can be classified as food with enhanced nutritious quality. This fruit belongs to Cucurbitaceae. This fruit can be rich in lycopene (Figure 7.1), carotenes, amino acids, vitamins A, B and C. Some of the nutritious elements that are present in these fruits can serve as an effective energy source. These elements can participate in the production of collagen. These elements can lower the oxidation rates. It also reduces the incidence of scurvy, improves cholesterol, and also inhibits lung, oral and stomach cancers. The fruits will get deteriorated because of the activity of the hydrolytic enzymes, and this can also lower the shelf life of the fruits. The main problem starts from the leakage of juices and soaking potential of the fruits.

Figure 7.1. Lycopene.

Source: https://en.wikipedia.org/wiki/Lycopene#/media/File:Lycopene_powder.jpg.

There will be a reduction of lycopene and soluble solids in the fruits. We can use the extracts from the Ocimum leaves (Figure 7.2) to preserve the freshly cut fruits. We can use bacteriocins to preserve the fruits. These compounds can be proteinaceous compounds. They can be produced as antimicrobial peptides. Most of these compounds can be ribosomally synthesized. The bacteriocins can be produced by most of the bacterial strains. These bacteriocins can suppress the growth of bacterial strains that are closely related species. In the direct approach, we can use bacteriocin directly into the food product of our choice. In the case of the indirect method, we can add lactic acid bacteria that can produce bacteriocins later as the food product ages.

Figure 7.2. Ocimum.

Source: https://en.wikipedia.org/wiki/Ocimum_tenuiflorum#/media/File:Ocimum_tenuiflorum2.jpg.

7.2. METHODOLOGIES

The methodologies that can be selected in the biopreservation process can be chosen based on both extrinsic and intrinsic factors that regulate the food products. Biopreservation based on the application of bacteriocins can be more productive. Probiotics can bring about the same changes, but there are certain limitations that are to be considered. Mostly bacteriocins can be preferred as they are not capable of changing the nature of the food products. Most of the physicochemical properties of the food products will be safe and never be disturbed due to the addition of the bacteriocins. Most of the bacterial species called as pathogenic bacteria will be suppressed, and this will enhance the shelf life of the food products. Guava fruits can be particularly important in most of the communities because of its medical values. This can be considered as a dietary component for common people who cannot afford costlier foods. They exhibit high nutritive value, aroma, sweet taste along with minerals and vitamins. We can also observe a certain proportion of carbohydrates in these fruits. Mostly, fruits constitute limited protein content and will have increased water quantity. The fruits are slightly acidic in nature, and this will allow the growth of yeasts, molds, and bacteria. Most of the Guava fruits are prone to wilt disease, and this will bring down most of the gains.

We can reap several advantages if we use a combination of techniques like UV irradiation, biopreservation and temperature regimes. These combinatorial techniques can work as a shield against several microbes, and this will extend the shelf life of these fruits. Most of the vegetables can be preserved using these techniques. Some of the bacteriocins can be produced by Bacillus cereus species. These compounds can be used to preserve the fruits for prolonged periods. Most of the bacteriocins can be safe for human consumption, and they can act against several microbes. Post-harvest, most of the nutritional, sensorial characteristics will be diminished with time. This can be attributed to microbial growth and food deterioration. These fruits can be susceptible to several attacks from the microbes, and hence there is a need for a series of postharvest treatments that can effectively lower the microbial loads and thereby extend the shelf life.

Because of improper postharvest management practices, 40% of the economy generated by the fruits will be lost. It is always important to check for quality postharvest and distribution. By providing membrane protection externally, lowering of microbial contamination, respiration and transpiration control can effectively reduce the losses incurred during postharvest stages.

It should be noted that the fruits that were minimally processed can be consumed by most of the customers as they are packed with nutrients. We can use agrochemicals for the effective growth of plants, and this is one of the reasons behind the onset of the fungal resistance by the microbes. Most of the chemical fungicides may not be useful to control the resistant fungal members. These fungal species can also produce toxic residues that can spoil the food products. The conventional preservation methods were deemed effective in preserving the fruits with limited advantages. We can opt for thermal processing even now to obtain desired results. In the case of thermal processing, we can use autoclaving temperatures that are heating at 121°C for 15 min. This technique can reduce the bacterial loads effectively. This process can be used on the fruits well before packaging, and it will be based on the fruit stuff, and it is not recommended. If the fruits have pH less than or equal to 4.6, we may need a temperature of 93° for 15 min to decontaminate according to the industrial standards. We can use chemical products that can reduce the microbial loads in the fruits after harvest. Some of the preservative products like ozone, chitosan, chlorinated water, hot water, mineral oils, ethanol can be used. The main problem with these compounds is the bitter taste and unfavorable odor. We can add chlorine to decontaminate the fruits. Sodium hypochlorite solution that is concentrated can be used in the preservation process. We can also use chlorine gas and calcium hypochlorite solutions. It is to be noted that the chlorine gas can be toxic in nature and should be used with caution. The increased amount of chlorine can damage skin and lungs, and it can also affect the coloration of the food products. Some of the equipment will be corroded.

Figure 7.3. Chlorine treatment.

Source: https://www.chemicalsafetyfacts.org/chlorine-dioxide/.

7.3. CHLORINE

We can use chlorine treatments (Figure 7.3) for the decontamination of fresh foods, and due to certain regulations and problems with chlorine, most of the

European countries have stopped using this process. It was reported that these halogen compounds can combine with the organic matter and produce toxic compounds that can be detrimental to human health. Whilst decontamination process, there will be production of harmful effluents. These disinfectants can produce bulk toxic compounds that can affect humans and animals in the vicinity. These products may not regulate the growth rate of the microbes, or they may not inactivate these microbes. It is all the concentration of these toxic compounds that matters as far as the decontamination is concerned. Utmost we can just obtain 2 log reduction in the load of the microbes. These concentrations may not act on the pathogenic microbes. The pathogens may need increased concentrations of toxicity. This is not possible with the concentrations that are being used to decontaminate the food products. The oxidation activity of the free chlorine can be affected by several factors. Some of the factors that can affect this oxidation activity are the presence of organic substances, pathogenic bacteria and its internalization and formation of biofilms. There will be presence of hydrophobicity in the plant surfaces, and that can also influence the action of free chlorine. It is always important to discuss the role of toxic substances for humans and the resistance that is generated in this process.

Figure 7.4. Organic acids.

Source: https://www.wbur.org/hereandnow/2020/09/16/canning-fruits-vegetables.

7.4. ORGANIC ACIDS

The organic acids (Figure 7.4) can play an important role in this process. The organic acids can effectively lower the microbial loads. This can be important in case of fruit preservation. We can use propionate and calcium lactate solutions to preserve apple slices and cantaloupes. In case of melons, we can use hydrogen peroxide solutions. Some of the organic acids that can be used in the preservation of the fruits are citric acid, ascorbic acid, tartaric, malic, and succinic acid. These organic acids can lower the microbial loads, and this can be attributed to the lowered pH levels. These organic acids can alter the membrane permeability, and this will effectively reduce the pH of the internal cell environment. Most of the metabolic enzymes will be affected, and protein synthesis will be impaired. Most of these organic acids can lower pH in several vegetables and fruits. There are several latest preservation techniques that are to be considered. Some of the preservation techniques are discontinued due to their involvement in the biochemistry of the fruits and vegetables. Some of the techniques that can alter the characteristics of the fruits and vegetables are ohm processing, pulsed electric fields, and microwaves. Commercial sterilization processes involve heating, and this can be called product sterilization. Due to increased commercial sterilization, there will be altered texture and color. Mostly, there will be loss of nutrients. Bleaching is another important technique that uses heating at least 70°C. This technique will reduce the microbial loads, and there will be inactivation of enzymes. This technique can yield drastic changes in texture and color of the fruits, and there will be loss of nutrients.

We can use chlorine in the decontamination process. This molecule can work as an oxidizing agent. The chlorination process is popular because it is cheap and limited concentrations are needed for the decontamination process. Some problems that are associated with the use of chlorine is its inability to destroy microbes in the fresh foods. Some of the equipment used in this process will be corroded, and there will be loss of color. Some of the microbes will become resistant to the pathogens. The organic acids are being used in the regulation of microbial growth. These organic acids can alter the genetic material of the microbes that are to be decontaminated. Most of the protein synthesis processes can be inhibited. Metabolic enzymes can be deactivated. There will be lowering of pH values. The organic acids can act on a broad spectrum of microbes. Due to the addition of the organic acids, we can expect bitter taste and some of the undesirable odors will be observed. The important techniques that are considered in the decontamination process are pulsed electric fields, microwave and ohmic methods. These methods can ensure uniformity and all the food products can be exposed to the same rate of decontamination. In these techniques, there will be rapid exposure of the food products within stipulated time intervals. The steam will be generated in the food products, and there will be conduction of heat all over the food products. We can observe local heating and this will lyse the microbial cells.

7.5. OHMIC HEATING

The ohmic heating (Figure 7.5) can be called the joule effect. In this process, foods will be heated, and this is possible with the application of the electric current at low frequency. The liquids and particles present in the food products can be heated at rapid speeds, but uniformity will be maintained. Some of the cold spots that can be present in the bulk food products will be removed. Post-heating, the food products will be cooled. After cooling, these food products will be packed carefully in the aseptic conditions. The microwave techniques can be used to destroy the pathogens and toxins. The food products will be sterilized, and minimal nutrient loss will be observed. This is far better than the conventional methods. The quality of the food products will be deteriorated.

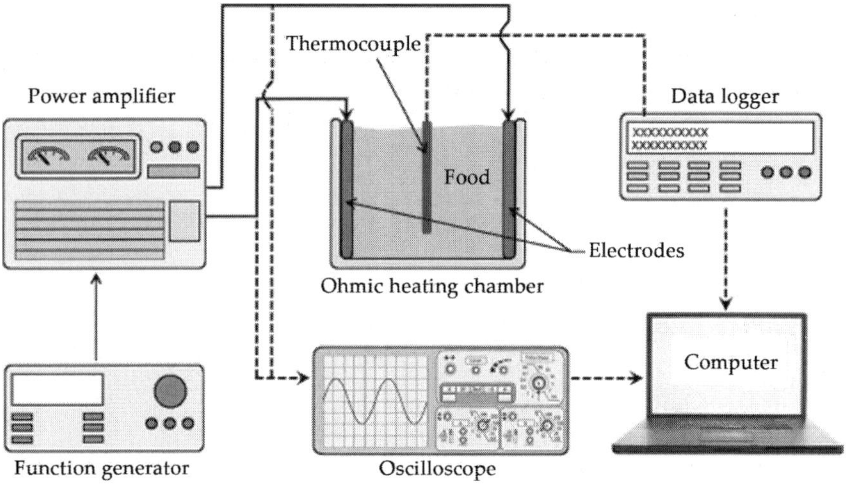

Figure 7.5. Ohmic heating.

Source: https://foodscience-technology.com/ohmic-heating/.

It is always important to ensure that the heating process is uniform throughout the foods. The pH of the foods can also affect the preservation process. We can use 915 MHz to enhance uniformity in the decontamination process. The pulsed electric fields can be used in the decontamination of the fruit juices. This will deactivate enzymes and destroy microbes. We can apply intense electric fields and pulses will be supplied in a regular manner. Most of the electric fields can be used to eliminate the microbes. The shape of the cells can be considered as an important factor to know the extent to which we can apply pulsed fields.

We can use UV radiation that can be nonionizing in nature. This radiation can be used in many applications. We can decontaminate food containers, surfaces, and water. Mostly, the surface of the foods can be decontaminated effectively. The main benefit of this method is that there will be no residues, and it can kill most of the microbes. We can gauge the efficacy of this method based on the equipment that is being used for this purpose. The turbidity that can be present in the liquids can halt the activity of UV radiation, and this can protect microbes from being degraded. We can also use high pressure preservation techniques to decontaminate the food products. In this method, there will be cleavage of the microbial membranes. In this method, we can use high pressures that may sometimes reach 8000 bars. At this pressure, most of the enzymes will be inactivated. These pressures can also inactive several microbes. The flavor and nutrients of the food products will be preserved. If the food products are porous, then the extreme pressures can damage their native texture. It is to be noted that the air molecules that are arrested inside the crevices of the food products will be altered on application of high pressures, and this can bring down quality.

The essential oils can also preserve the fruits or fruit-based products. Some of the chemical preservatives that can be used in the storage of the food materials can produce allergic reactions in humans. We can get these essential oils from spices and herbs. These oils are meant to be used in the food products as aroma producing agents. They can be highly volatile, and most of them will be produced due to secondary metabolism, preferably in plants. They can have an excellent complex mixture of phenolic acids, terpenes, ketones, aldehydes, and esters. These compounds can be called bioactive compounds. Some of these compounds can be aliphatic and aromatic in nature. Most of these essential oils can be used to regulate the extent of food spoilage. Some of the pathogenic bacteria can be eliminated, and this can also halt lipid oxidation.

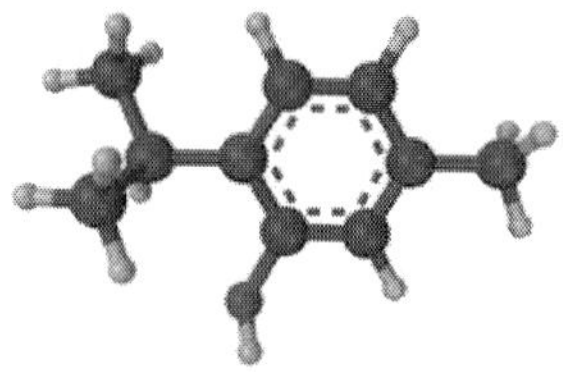

Figure 7.6. Thymol.

Source: https://en.wikipedia.org/wiki/Thymol.

Thymol (Figure 7.6) is present in these essential oils and it can kill most of the microbes. Thymol can be produced from oregano and thyme. We can extract eugenol from the cloves generally. Cinnamon (Figure 7.7) can be used to isolate cinnamaldehyde. In case of ohmic preservation, we can use alternating current at 60 Hz. In this technique, we can expect uniform heating, and this will be rapid. In the case of microwaves (915 MHz), we can expect irregular heating and this can be attributed to the pH values. Electric pulsed fields can be used, and it includes application of intense electric fields at short pulse frequencies. These pulses can kill microbes. Due to this technique, there will be damage to the cell, and this can be based on the structure of the tissues. We need to consider the shape of the tissues also. The electrophysical properties of the cell must be defined properly. The ultraviolet radiation at the range from 200 to 280 nanometers can be used in the sterilization process. In this technique, there will be no residues. The optical properties of the foods and flow patterns along with turbidity can also affect the action of UV radiation in the decontamination process.

Figure 7.7. Cinnamon.

Source: https://www.emedihealth.com/cinnamon.html.

The high pressures that range from 3000 to 8000 bars can be used in the decontamination of the food products. This can be based on the integrity of the food products. The porous flesh containing food products can be mostly affected by the high-pressure treatments. We can use essential oils as mentioned above to decontaminate the food products effectively. These oils can degrade the proton motive forces. There will be coagulation reactions in the cytoplasm. The membrane proteins and cytoplasmic membranes can get damaged. Some of the portions of cell walls can be degraded. These oils can also regulate the spoilage of the foods. Some of the organoleptic properties of the food products can be altered. We can expect strong aroma changes in the food products that were preserved with essential oils. Coatings and edible oils are important methods that can be used in the preservation process. There will be a reduction of senescence and respiration. These coatings can be cheaper and mostly non-toxic in nature. In stored conditions, we can maintain quality for prolonged periods. Some of the polysaccharides that are being used in case of coatings and edible films cannot tackle moisture. Most of the proteins may exhibit limited resistance against water vapor. Some of the proteins can easily get degraded by the action of proteases. We can use modified atmosphere packaging methods in the preservation process. The activity of the product will be reduced, and this can be attributed to the reduced respiration rates. There will be inhibition of several microbes, and hence most of the fruits can be saved for prolonged periods without spoilage. The balance that exists inside the fruits will be affected in this method. We can find it difficult to orchestrate the exact combination of gases that are to be used in this process. The respiration reactions of the food products can also alter the proportion of gases that might exist in this method. This system or technique can be classified as a nonsterile system. We can also opt for active packaging systems. There will be liberation of antimicrobials in the package. With this technique, there will be quality intact and extended shelf life. In this technique, the food should be in direct contact with the packaging material. Control of oxygen can be harder to achieve. Bioprotection techniques can be used in this process. There will be induction of the antimicrobials. Salinity can be introduced in this method. We need to consider moisture dynamics in this method, and there will be a lowering of pH values. Some of the metabolites will be released into the food products. We can easily eliminate the pathogens as the niche that exists in the food microenvironments will be disturbed. This technique is not suitable for vegetables.

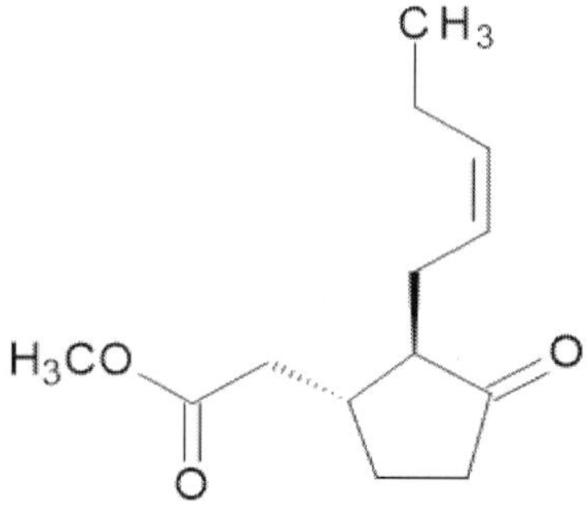

Figure 7.8. Methyl jasmonate.

Source: https://www.emedihealth.com/cinnamon.html.

In the essential oils that are extracted from jasmine, we can observe the presence of methyl jasmonate (Figure 7.8). Mostly, the essential oils that are extracted from the plants can induce several defense mechanisms that can kill several microbes, and some of the pests can be killed also. This oil can be antifungal and antimicrobial in nature, and hence most of the bacterial and fungal species will be destroyed. The essential oils induce antimicrobial activity in a series of steps. There will be participation of diverse chemical compounds that can take over several mechanisms. Some of the internal mechanisms and membrane proteins will be affected by this essential oil. The essential oil produced from Basil can be used to regulate gram positive bacteria. Some of the bacterial species that can be controlled by the extracts produced from basil are Bacillus subtilis, Staphylococcus aureus, Staphylococcus epidermidis, Sarcina lutea, and Micrococcus flavus. This oil is also effective against gram negative bacteria. Some of the bacterial species that can be regulated are Shigella sonei, Salmonella enteritidis, Salmonella typhi, Pseudomonas aeruginosa, and Escherichia coli. Most of the gram-positive bacteria will be controlled on application of essential oils that are extracted from basil. These oils can induce strong aroma and that will affect the palatability of the food products. Some of the sensory characteristics can be affected due to usage of these compounds. This is the reason why this oil may not be regularly used in food preservation. There are several assessments that were attempted on the essential oil and their role in preserving the food products. The coatings and edible films can be used in extending the shelf life of the food products like fruits. They are recommended because of their low cost, limited environmental impact, nontoxicity, and biocompatibility. These compounds or formulations can be used in maintaining the food quality.

7.6. EDIBLE COATINGS

These coatings are edible, and they can be used in the form of food grade lipids and hydrocolloids. Mostly, we can apply these coatings onto the food surface easily. This can be done by spraying. We can also use dipping or smearing techniques also. Once the applied material dries, we can observe thin layers on the food products. The coatings thus applied can reduce the respiration rates of the food products. These layers can act as a barrier for carbon dioxide, oxygen, and water vapor. This will result in the increased humidity inside the food products. Most of the redox reactions will be slowed down. The aroma of the food products will be saved along with its color and texture. The main advantage with these coatings is that there will be negligible microbial contamination. These films are mostly edible, and they can be used as vectors to supply antimicrobial agents. These films can also get into final products. These coatings can be produced using four basic materials. They are proteins, polysaccharides, resins, and lipids. Most of the polysaccharides can generate edible films that are homogeneous in nature and they are transparent. These coatings can exhibit selective permeability against carbon dioxide and oxygen molecules. These films are resistant to both oils and fats. This can be attributed to the structured network that is ordered and there will be H bonds that are densely packed. They do not have taste and odor, and this makes them an ideal candidate. These coatings can be hydrophilic in nature and hence they may not work well against moisture. Some of the edible films can be produced by using proteins. These films can be edible and act as a barrier. This barrier can be effective against carbon dioxide and oxygen. The application of these films can be limited as there will be limited resistance against water. This can be attributed to the presence of proteolytic enzymes and proteins. The coatings that are produced with the aid of lipids can work effectively against moisture. Though they can change the appearance of the food products, there will be certain limitations. We can enhance the strength and function of these films with the use of additives. Some of the additives that can enhance the quality of the food products and preservation can be essential oils, organic acids, antioxidants, antimicrobial compounds, etc. We need to evaluate the coating components and additives before using these films. These coatings can also enhance the efficacy of preservation and can act as fungicides.

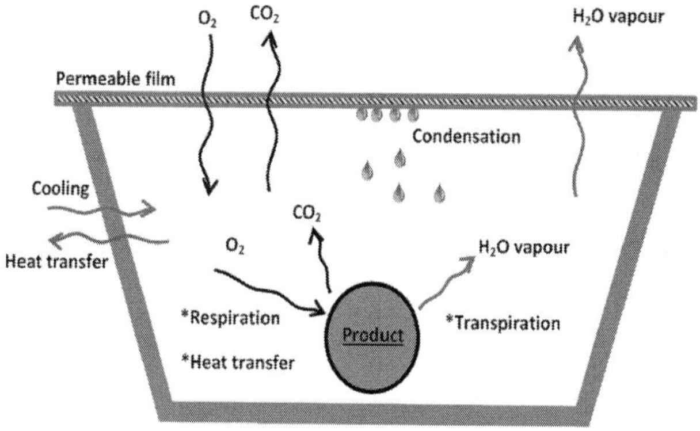

Overview of the continuous phenomena (*) in modified atmosphere packaged fresh product.

Figure 7.9. Modified atmosphere packaging.

Source: https://ars.els-cdn.com/content/image/1-s2.0-S2214289416300813-fx1_lrg.jpg.

The modified atmosphere packaging (Figure 7.9) can also lower the microbial growth rates in the food products that are fresh. Most of the fresh foods can be preserved using this technique. In this technique, we can use differential composition methods to moderate the atmospheric conditions inside. Whilst transit of these food products, the proportion of gases and storage temperature can be regulated by altering the oxygen supply ratio. We can also supply at least 5% of carbon dioxide into this system. In this method, there will be reduced microbial growth and respiration rates. The gas proportions that can be used in this technique can be determined well in advance prior to food preservation. We need to consider pH, water activity, natural microbiota, constituents, composition, and maturity levels of the food products. Most of the fruits can be demarcated by their native microbiota, and there can be fungi and bacterial species. These species can enhance the food spoilage rates. Most of the pathogens that are present in the food products should compete with the microbial species that may attack in storage conditions. The microenvironment in the stored food products can alter all these components, and this will lead to a selective approach. Due to a reduced oxygen environment, most of the aerobic microbiota will be inhibited. These oxygen limiting conditions can also promote the development of other bacterial species that can grow in limited oxygenated conditions.

Active packing can be another important technique that can be used in the preservation of the food products. In this technique, the fruits that are stored will be interacted with the active packaging cascades. This will extend the shelf life of the fruits. Bioprotection can be used to preserve the fruits, and this can extend the shelf life of the fruits and hence food safety can be ensured. We can use several beneficial microbes that can regulate spoilage specific microbes that may deteriorate the food products. There are some microbes that can produce metabolites that can easily regulate the pathogenic microbes. The generation of the microbes can be considered as a part of natural equilibrium. We need to determine the essential features of the microbes that can be used to preserve the food products. This is necessary because we cannot use microbes that can create additional problems in the food products. The sensorial and nutritional properties should be saved in the food products, and this will also encourage customer satisfaction. Some of the pathogens that are present in the food-based systems can lead to several fluctuations in the microenvironment of the foods. We need to determine the microbial communities that may survive in the food products even before adding preservatives that are natural. Some of the pathogens may be killed, and this can be possible by minor alteration in the niche.

In this process, we can use lactic acid bacteria with several benefits. This bacterium can minimize the colonization of the bacterial species that can be pathogenic in nature in normal conditions. The lactic acid bacteria can exhibit antagonistic activity, and this can be severe against several pathogens. These bacterial species can be classified as GRAS. We can produce several bioprotective agents that can preserve vegetables and fruits. We need to select and isolate important biocontrol agents from the native environment that might exist in the food products at a specific period. Most of the metabolites that were produced from the lactic acid bacteria are antimicrobial in nature. Some of the metabolites are bacteriocins, ethanol, lactic acid, and other compounds. These bacteria can grow in the cold temperatures and can be used as preservatives. We can use purified metabolites and fermentation products in place of these preservatives. We can use this bacterial species even to store the nondairy sources. We can use these cultures to preserve vegetables, fruits, and cereals. We can easily isolate the lactic acid bacterial strains from pickles that were prepared using traditional techniques. We can collect it from rural areas and ensure that appropriate bacterial strains are collected. Some of the Shigella dysenteriae, Staphylococcus aureus, Escherichia coli, and Bacillus cereus can be isolated from these pickles. From the corn products and fermented milk, we can isolate these species. The metabolites generated from Lactobacillus plantarum can exhibit antifungal activity. Aspergillus flavus can be effectively controlled due to the addition of Lactobacillus plantarum. The lactic acid bacteria or LAB can be safely isolated from the lettuce and apples. This species can effectively destroy the pathogens that can spoil the food products. The bacteriocins that were developed from L. pentosus, W. soli, L. mesenteroides, and L. plantarum can destroy L. monocytogenes, E. coli, and S. enteritidis. Lactic acid bacteria can lower the load of aerobic bacteria that are mesophilic in nature. This action can be the same as the action of chlorine. We can isolate LAB strains from the fermented products and fresh vegetables. These strains can effectively reduce the microbial growth rate of Weissella cibaria, Lactobacillus plantarum, Lactobacillus fermentum, and Staphylococcus aureus. Pediococcus pentosaceus, Pediococcus acidilactici, and Lactobacillus sakei can exhibit antimicrobial activity against yeasts, fungi, and bacterial species. The metabolites that can be produced from the lactic acid bacteria can reduce the growth of Escherichia, Listeria, Pseudomonas and Bacillus species. Most of the metabolites generated from these bacterial species can exhibit fungistatic and fungicidal activity.

The Enterococcus faecium, Weissella halotoletans and Leuconostoc mesenteroides can be isolated and cultured from the rhizosphere layers of the desert truffles and olive trees. These species can attack the pathogenic microbes like fungi and bacteria. Some of the foodborne pathogens like Listeria monocytogenes and Staphylococcus aureus can be regulated in addition to these bacterial species. Some of the lactic acid bacterial species can be produced from the olives, and these species can produce metabolites that can act against Salmonella, Bacillus cereus, and Listeria monocytogenes. The lactic acid bacteria species can be isolated from vegetables, flowers, and fruits. They can be isolated from the fermented products.

L.brevis strain can produce brevicin. The bacterial growth can be reduced on application of these bacteriocin producing strains on the iceberg lettuce leaves and apple lesions. These strains can be easily cultured on the suitable substrates. The bacterial species like E. coli and S. typhimurium can be inhibited. Due to increased antifungal activities, there will be minimal production of aflatoxins. The bacterial species like Lactobacillus fermentum and Bifidobacterium bifidum can act against Aspergillus parasiticus species. Most of the metabolites have reduced the level of aflatoxins production. Colletotrichum capsica can be regulated by adding L. plantarum and L. paracasei species. These species can act as biocontrol agents. This fungal species can affect chili peppers by causing anthracnose. The supernatants have exhibited excellent inhibitory activity on this fungal species.

There are some antimicrobial effects exhibited by these bacterial species. We can also use diacetyl compounds, hydrogen peroxide, and organic acids to regulate the microbial growths. Some of these compounds can behave like peptides and proteins. The peptides that are derived from the lactic acid bacteria can kill several pathogens and can also inhibit certain species that can be resistant to the antimicrobials generated from other species. P. pentosaceus. L. plantarum, and E. fecalis can act on pathogens like S. aureus, E. coli and B. cereus. The Enterococcus sp., L. lactis and L. mesenteroides can act against S. dysenteriae species that can be pathogenic in nature. L. plantarum can inhibit the growth of A. flavus. L.f ermentum and B. bifidum can act against Aflatoxins and A. parasiticus. L. plantarum, W. soli, and Lm. mesenteroides can inhibit the pathogenic species like E. coli, L. monocytogenes and S. enteritidis. The L. pseudomesenteroides, W. cibaria, E. mundtii, P. pentosaceus and L. plantarum can attack pathogenic species like E. coli, M. luteus and S. enterica. We can isolate Weissella cibaria, Lactobacillus plantarum and Lactobacillus fermentum species from fresh vegetables and these species can inhibit the growth of S. aureus. P. pentosaceus, P. acidilactici and L. sakei can inhibit the pathogen species like C. parapsilosis, D. hansenii, A. niger, P. expansum, A. versicolor, A. fumigatus, P. chrysogenum, F. culmorum, Listeria, Pseudomonas, and Bacillus species.

There are several benefits of the lactic acid bacterial species. This can enhance the durability of the food products. This can be attributed to the decreased water activity and production of the metabolites. On application of LAB, there will be enhanced sensorial and nutritional values. Microencapsulation technique can be used in the preservation of the fruits. We need to determine the viability range of LAB species. This can be important while dealing with the bioprotective cultures. There are several methods that can be used to enhance the viability of the cells. Most of the LAB cells can be immobilized. This process can be done through a droplet matrix. This can be assumed as a hydrocolloid. There will be increased protection for these food products from the environmental conditions. We need to enclose the particles that are in the range of micrometers. These particles can be gases, liquids, droplets, and solids. These entities can be packed into a capsule. There will be the formation of an envelope that is thin, and we can define it as an encapsulant. These encapsulants can enhance protection against diverse environmental stresses. Some of the environmental stress can be in the form of humidity, light, air, and heat. These encapsulants can work against predators and competitors, and this can be effective in the refrigeration conditions also. We need to add these encapsulated microbes such that they survive for prolonged periods. The LAB species can exert biological effect and there will be enhanced activity. It can be important in the form of probiotics, and mostly it gets settled in the colon after reaching the digestive tract. Emulsions can be classified as important techniques that can be studied under microencapsulation technology. The emulsion of LAB can be simple and economical too. We can use emulsions in the large-scale production of the preservatives. We can encapsulate emulsions using the hydrogel precursor dispersal method. This can be done using immiscible phases. Most of these emulsions can be stabilized with the aid of surfactants. In this emulsification method, we can use a batch production process. It needs more steps to eliminate the surfactants that are added in the process.

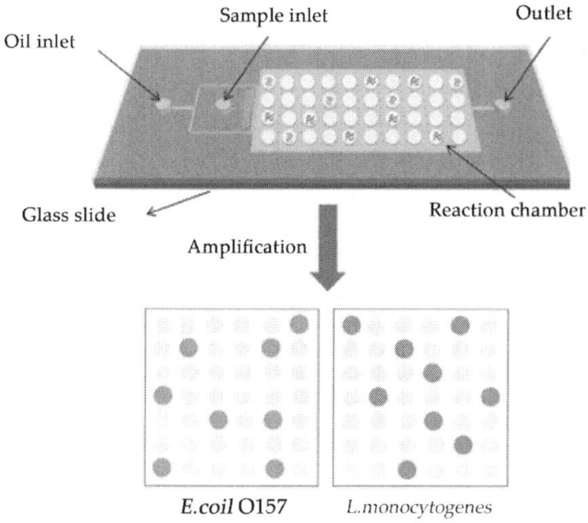

Figure 7.10. Microfluidics.

Source: https://www.sciencedirect.com/science/article/pii/B9780128142172000275.

7.7. MICROFLUIDICS

Microfluidics (Figure 7.10) can be an important technique in the encapsulation process. The microfluidics can be used in transferring the fluids in microchannels. This process can occur via laminar flow. The microdroplets can be produced. There will be the production of emulsions in the form of droplets that are made up of biopolymers. There will be establishment of the immiscible phase that can be continuous in nature. We can use spray techniques in order to generate minute microcapsules, and this will occur at enhanced speeds. The coating materials can be considered as important technique in the encapsulation process. There should be chemical compatibility. The encapsulating component must be inert to other compounds. There should be a certain range of affordability so that other food-based companies can use this technique. These capsules can enhance stability, permeability, flexibility, and strength. We can use matrices in the microencapsulation process, and this can be done with alginate members. It can be nontoxic in nature. There will be limited immunogenicity. We can expect minimal gelation in this process. These substances can be called as GRAS. These emulsions can act as bioactive compounds that are naturally derived. We can use alginate easily and this can prolong the survival of useful bacteria. The increase in the alginate concentrations can also improve the colony-forming units. This can be observed in case of Bifidobacterium longum species. The microparticles made up of alginate can be seldom chemically stable in nature. Most of the encapsulated materials will be protected. This can be attributed to the presence of the cations that may not form gels. There will be certain cations like magnesium and sodium along with chelating agents like citrate and phosphate that can be used in this process. We can use co-encapsulating compounds that can enhance the strength of the alginate matrix. We can cover these capsules with the aid of cationic polymers. We can enhance the stability of the microparticles. The materials that are encapsulated minimally can be susceptible to the damage patterns. We can opt for controlled release of the LAB species in the food products. By spray and freeze-drying methods, we can encapsulate cultures of L. plantarum, B. amyloliquefaciens, E. fecalis and E. hirae species and skimmed milk can act as a good substance for matrices. Through fluid bed drying and freeze-drying, we can encapsulate L. plantarum with chitosan and alginate molecules. The extrusion process can be used in the encapsulation of L. curvatus based on calcium alginate. Ionic external gelation process can be used with chitosan, tween 80, calcium chloride and sodium alginate in the encapsulation of L. bulgaricus, L. acidophilus and L. plantarum. We can use this method for encapsulating bacterial species via Arabic gum, alginate, gellan gum also. Emulsion can be used to encapsulate B. bifidum and L. acidophilus using Tween 80 and Carragenate corn oil. We can use both spray drying and emulsion to immobilize the B. infantis. In this process, we can use canola oil with water. We can also use micro fluidized resistant starch and dried glucose, caseinate in this process.

We can use bacteriocins in this process, and most of these compounds can be antimicrobial in nature. Lactic acid bacteria and Bacillus species can be used in the production of bacteriocins. We can use these chemicals as biopreservatives. We can incorporate these chemicals in the coatings that are edible. Garden strawberries can be classified as important fruits that are to be preserved, and this can be due to demand from the public. This fruit is one of the important components in the human diet. We can use this fruit in processed or fresh forms. This can be due to its unique flavor and taste. Most of these fruits can act as strong antioxidants. This can be attributed to the presence of ascorbic acid, carotenoids, tocopherols, nitrogenous compounds, flavonoids, phenolic compounds, and anthocyanins. The strawberries can be nutritious in nature and it is healthy for humans. This fruit also contains sulfur compounds, alcohols, aldehydes, and esters that can be called volatile compounds. Due to these compounds, these fruits exhibit increased antioxidant capabilities. There will be peculiar phytochemical composition in these fruits that can be beneficial for human nutrition. These compounds can be active in treating cardiovascular diseases and other degenerative diseases. These compounds can act as both anticarcinogenic and antimutagenic in nature.

These fruits can be perishable and sensitive, and this can explain the minimal shelf life of the fruits. Most of the infections that damage these fruits can be due to physical injuries. There will be the appearance of several disorders and other conditions due to fungal induced decay. The postharvest decay of strawberries can be prevented by storing them at very low temperatures at least for 4 days. The optimum temperatures at which we can store this fruit is at 0 to 4°C. We can use edible coatings to preserve this fruit. We can use these coatings in the form of thin layers on the surface of the fruits. These coatings can be considered as the best option to save strawberries for prolonged periods. These coatings can safeguard the microbial activity, physicochemical activity and keep the mechanical damage at minimum. The fruits will not be dehydrated as these coatings can act as barriers to stop the moisture from being lost. This can be considered as the main reason why fruits will get spoiled. The stiffness and mass of the fruits will be reduced. The quality of texture will be diminished, and this can be attributed to the reduction of the respiration rates in fruits because of the edible coatings. The flavoring compounds are mostly volatile and most of the natural colors will be sustained even after storage. This may lead to diminished fungal growth inside and on the surface of the fruits. The selection of materials that are needed in the preparation of coatings can be a crucial task. We can use combinations or a single material to prepare coatings. The edible coatings can be prepared from lipids, proteins, and polysaccharides, and in some cases, we can use their derivatives. These compounds can be used in the form of matrices. These matrices can be called structural matrices. We can use cellulose molecules as they are naturally available polysaccharides. We can use carboxymethyl cellulose, hydroxypropyl methylcellulose, methylcellulose, and hydroxypropyl cellulose can be used as biopolymers that can be used to preserve the fruits. These compounds can be injected into the edible coatings that are commonly perishable in nature. Most of the derivatives of cellulose can be water-soluble, tasteless, and in some cases odorless. These compounds can exhibit film forming capabilities. These films can form a barrier in transmitting oxygen and moisture. The carboxymethyl cellulose can be used as raw material in most of the cases as it has many advantages. We can use pediocin, bificin C6165, enterocins 416K1 and As-48, bovicin CH5 and nisin in the biopreservation of the fruits.

Chapter 8

Biopreservation of Milk Products

CONTENTS

8.1. INTRODUCTION

Most of the latest technologies have become part of the food preservation industries, and these technologies can ensure food safety standards. This has reduced several foods linked illness, and most of the food spoilage will be reduced due to intervention of these techniques. There has been excess consumption of the precooked foods in recent years. This is considered as unsafe as some of the products were not stored at optimal temperatures, and this condition is common in the developing countries. Most of the foods will be contaminated due to the action of pathogens. This is the main problem with the foods that are stored in irregular temperature intervals. It is to be noted that at least 350 million people are affected with gastroenteritis due to food borne illness. Most of these cases can be attributed to the food borne pathogens. Some of the main pathogens that can cause food borne illness are Staphylococcus aureus, Listeria monocytogenes, Escherichia coli, Campylobacter jejuni, and Salmonella species. There are certain traditional methods that are in place to preserve the milk products, and some are upgraded according to the customer demands. There were several methods that are being attempted in the biopreservation of the milk products. With the biopreservation methods, we can easily extend the shelf life of these products. The hygienic quality of these foods will be enhanced. There will be improved organoleptic properties and nutritional qualities of the milk products (Figure 8.1) that are perishable in nature.

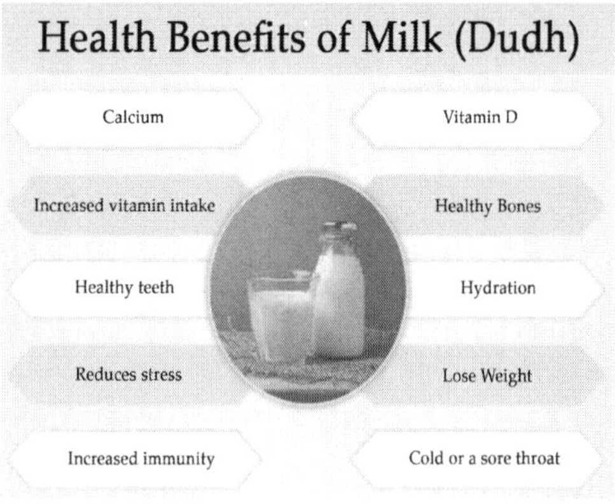

Figure 8.1. Nutritional qualities of the milk.

Source: https://www.lybrate.com/topic/milk-dudh-benefits-and-side-effects.

Biopreservation technique uses antimicrobial qualities that can inhibit pathogens and other microbial species that may spoil the food products. This process can enhance the safety levels of the food products. Biopreservation can be a technique that can use microbiota that can be natural and, in some cases, controlled to preserve food products along with extending their shelf life. We can also use antimicrobial compounds that can diminish the growth of harmful bacteria. Fermentation is another important process that can be classified under biopreservation. This process can leave the complex compounds into simpler compounds. There will be production of vitamin-C precursors, riboflavin, vitamin-B12, and other compounds that can act as antifungal agents (Figure 8.2).

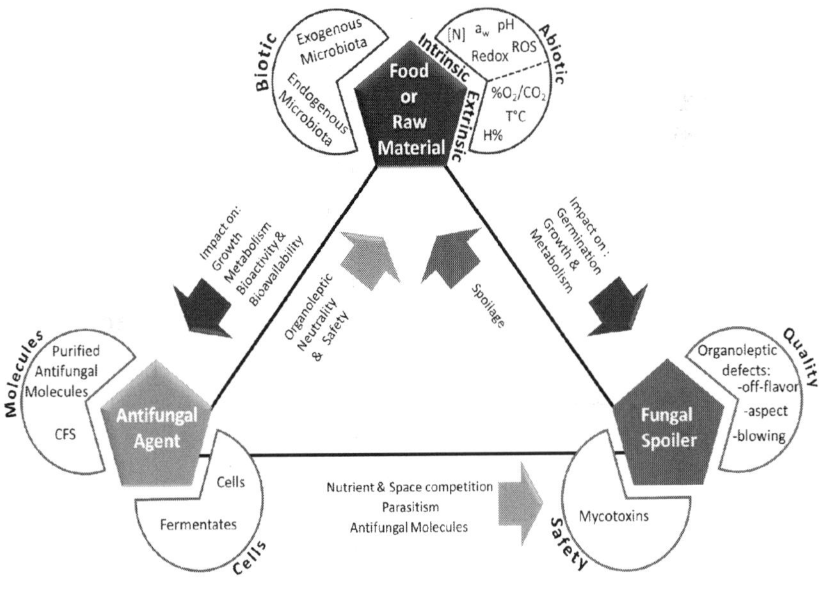

Figure 8.2. Antifungal agents.

Source: https://www.mdpi.com/2076-2607/5/3/37/htm#.

Due to the production of these compounds, we can enhance the organoleptic properties of the food products. There will be production of aroma and flavor in these compounds due to biopreservation. These organisms can be dominated by lactic acid bacteria. There will be the production of organic acids and other essential compounds that can exhibit antimicrobial activity. There will be the appearance of textures and flavors in these food products. Several food products were protected against the bacterial infections, and this can be attributed to the fermentation process.

These fermented foods have gained high priority these days. Most of the developing countries are using these fermented foods in their diet needs, and this has worked out well in the form of probiotics. The food-based industries in the developed countries are aiming at the use of protective cultures. The enhanced organoleptic properties are attributed to the microbial diversity in the fermented foods. All these beneficial properties can be attributed to the microbiota.

There are several methods that can be used in the biopreservation process. We can use starter cultures in this process. The starter cultures can be extracted from the fermented foods. These starter cultures can be used to initiate the fermentation process. This process can be adapted by several food industries. Mostly, dairy industries can be based on the fermentation systems. We can use the fermentation process in the preparation of juices, vegetable products, and meats. We need to select bacteria based on the type of food product. These microbes can improve the food products in terms of biological, chemical, and physical properties. Due to improvement in the components in the food products, we can expect increased flavors. In order to enhance the food products using microbes, we need to get approval from FDA, and these microbes should be classified as generally regarded as safe or GRAS.

These microbes should not be toxigenic and pathogenic in nature. We need to ensure that these microbes can be reproducible and can be standardized. We can use lactic acid bacteria in the biopreservation process. The LAB species can be used as natural biopreservatives. They can exhibit many antagonistic activities. Most of the pathogens and spoilage bacteria can be inhibited by the LAB species.

8.2. LAB

The LAB derived metabolites can contain certain components that can be classified as acid components. Some of the acids that are present in these metabolites are bacteriocins, hydrogen peroxide, lactic acid, and acetic acid. The lactic acid bacteria can compete with other microbes for nutrients, and due to this, there will be production of metabolites. This bacterial species can act as a perfect preservative for milk-based products. We can use bacteriocins that were produced by LAB species with the aid of hurdle technology. We can use bacteriocins as an effective preservation technique that can regulate the microbes that spoil the food products. It also inhibits bioactivity, growth, and generation of the microbes.

Most of the bacteriocins can be produced in the form of peptides or even as proteins. These compounds can exhibit antimicrobial properties that can kill several microbes. These compounds can be produced from the ribosomes of bacteria. These compounds can be classified as bacteriocins, and most of them are produced from the LAB species. These peptides can kill most of the bacterial species that are associated with LAB. These compounds can withstand heat and they are smaller in size.

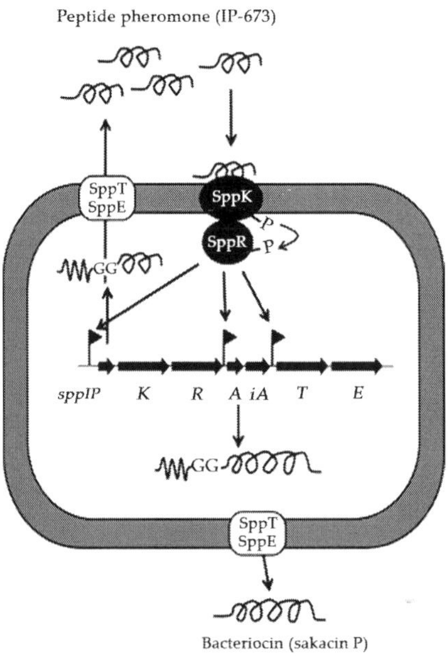

Figure 8.3. Sakacin.

Source: https://www.sciencedirect.com/topics/agricultural-and-biological-sciences/sakacins.

These compounds can be classified as cationic peptides. Most of these compounds can be produced by gram-positive bacteria. Lactin and Nisin can be used along with food products, and they can act well against mastitis. We can use these bacteriocins instead of antibiotics in the industries to regulate pathogenic cultures. The LAB species can easily colonize the gastrointestinal tract of humans and many animals. Some of the zoonotic pathogens can be minimized. These compounds can be used to speed up the ripening process in cheese. Due to application of these microbes, there will be activity of the intracellular enzymes so that the matrix of the cheese will be altered. The volatiles can be heavily concentrated, and this is the reason why most of the cheese products can develop specific sensory attributes. The matured cheese can be prepared by using the bacteriocins. We need to inoculate the LAB cultures or bacteriocins in the food products to yield benefits. We can use both semi-purified and purified bacteriocins in this process. We can use nisin producing strains or nisin itself to control clostridium species that can be present in the cheese products. We can also use these compounds to control Listeria species that attack camembert.

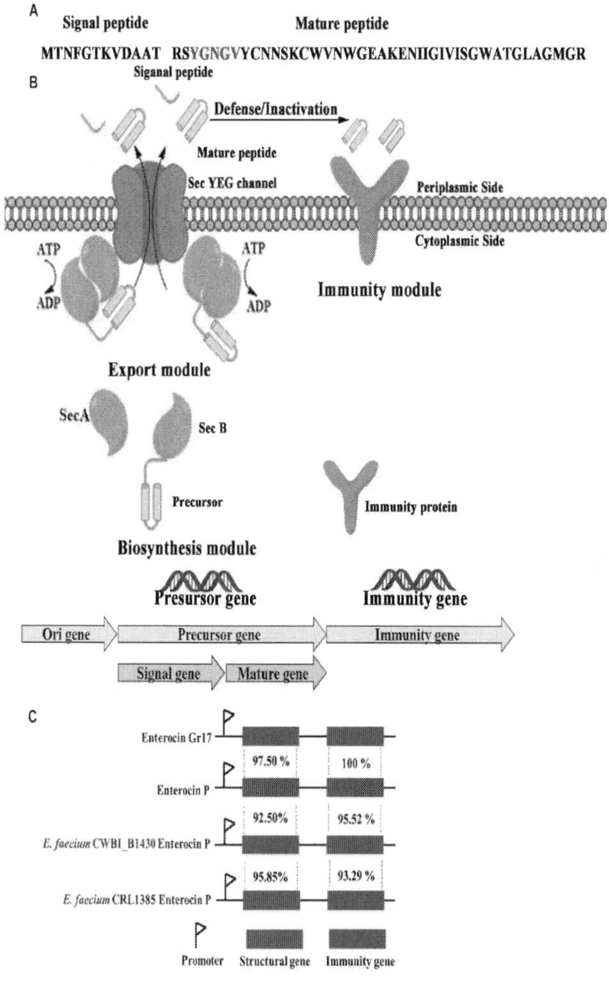

Figure 8.4. Enterocin.

Source: https://www.frontiersin.org/files/Articles/435658/fmicb-10-01806-HTML/image_m/fmicb-10-01806-g002.jpg.

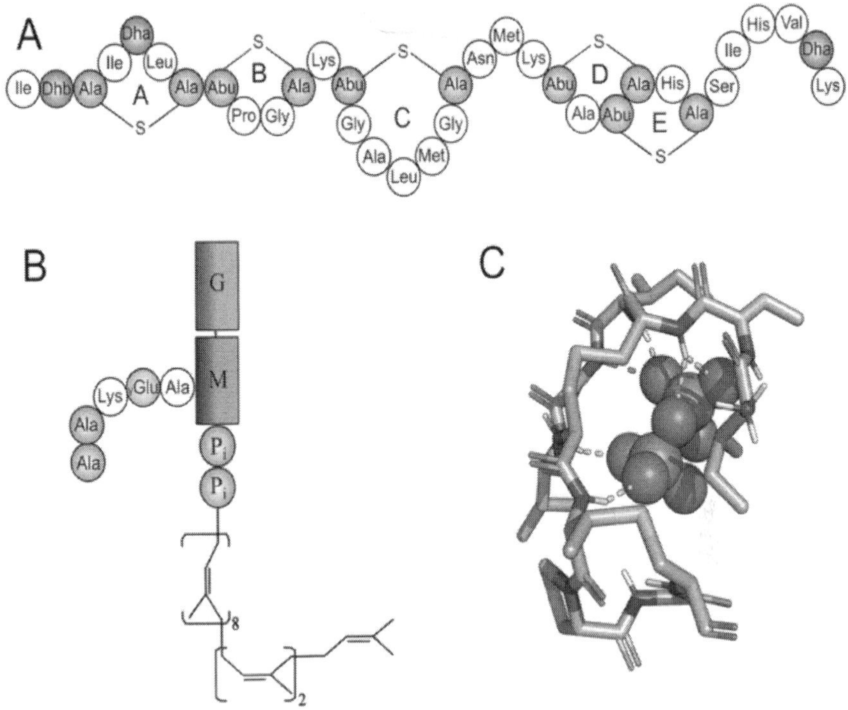

Figure 8.5. Nisin.

Source: https://www.nature.com/articles/s41598-020-65522-y.

We can use sakacin (Figure 8.3), enterocin (Figure 8.4) and nisin (Figure 8.5) also. Most of the fish products can be affected by the gram-negative microbes. This can be common in vacuum packed seafoods and fishes. We can observe Listeria and Clostridium species in the packed foods. We can use both Microgard and nisin to lower the load of aerobic bacteria. This can be done in case of salmon that is freshly chilled. On application of these compounds, we can extend the shelf life of the fishes that are in storage. We can also control the growth of Listeria species even after thawing.

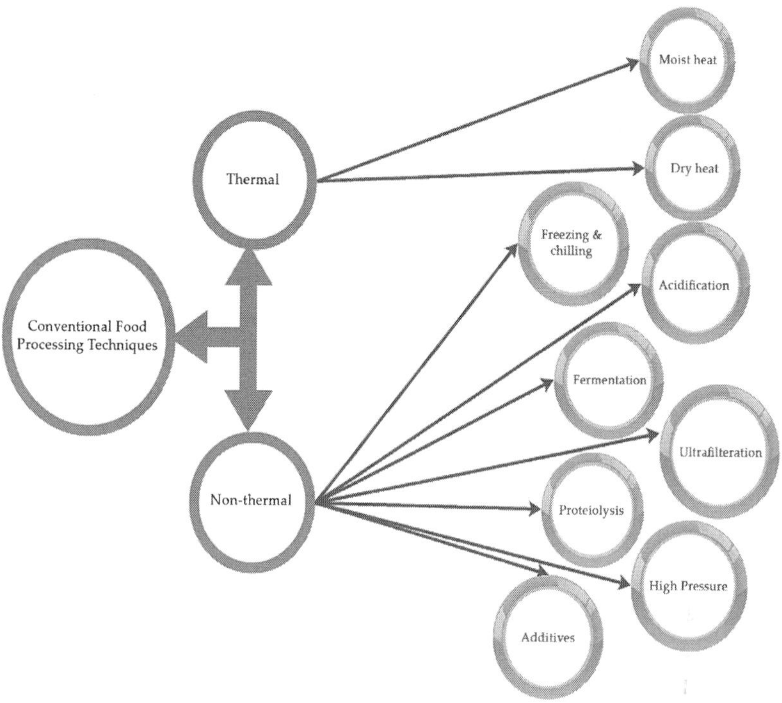

Figure 8.6. Hurdle technology.

Source: https://www.sciencedirect.com/science/article/abs/pii/S0956713516306247.

8.3. HURDLE TECHNOLOGY

Hurdle technology (Figure 8.6) can be used in this preservation process. This concept was materialized by Leistner in 1978. The hurdle concept was used to enhance nutritional, sensorial qualities of the food products along with the stability and safety. We can combine preservation techniques to preserve food products. These factors can be called hurdles. Most of the microbes that can spoil these food products may not become tolerant or resistant to these hurdles on a long run. Some of the hurdles that can be used in the biopreservation process can be redox potential, water activity, increased acidity, lowering of pH, storing at optimal temperature and food processing at increased temperatures. The tendency of the hurdles to inhibit the microbial growth must be defined, and it is to be ascertained based on the type of microbes that attack the food products. We need to consider preference and safety standards specified by the law enforcing agencies pertaining to the food products. These chemicals should not disturb the appearance of the food products. We can use hurdle technology to safeguard diverse food products. We need to consider standard protocols in the preservation of the food materials.

Hurdle technology is considered as a preferential treatment strategy that can be used to increase the shelf life of the foods. The foods can be subjected to hurdle technology, and this is preferred in the case of foods that were minimally processed. This can be used as a better solution in case of fat-enriched foods. We can apply this technology on the foods that are refrigerated and fermented, preferably at low temperatures. This can lead to intoxication of the food products due to alteration of optimal conditions of the food storage. In the developing countries, the food products can be preserved without any special refrigeration. This implies that there is a need for empirical utilization of hurdle technology. There are many foods that can be classified as traditional foods, and most of them were tested with this hurdle technology to know the actual mechanisms that govern the preservation process. Commonly used hurdles in the food preservation process are competitive flora, UV, HHP, modified atmosphere, vacuum packaging, chemical preservatives, redox potential, pH, water activity and temperature. This process is becoming popular these days due to its acceptability and uses.

The lactic acid bacteria can comprise a diverse group of microbes, and each microbe can exhibit diverse metabolic properties. The microbial properties can be attributed to the diversity in the microflora present in the food products. This is the main reason why the acid fermentation of food is most popular. The LAB species has limited taxonomic significance. This is defined by serological techniques. Phylogenetic studies were made on LAB species using rRNA (16S) cataloging technique. According to these studies, LAB species have shared characteristics with several ancestors. The dominant species in the lactic acid family are Pediococcus, Leuconostoc, Streptococcus, Lactobacillus, and Lactococcus. There are several species that were used in the biopreservation process but mostly limited positive results were yielded. Most of the lactic acid bacteria can produce antibiotic peptides and they comprise 24 amino acids. We can specify certain antibiotic metabolites as lantibiotics. This is due to the presence of Lan or Lanthionine amino acid after translation. BLIS or bacteriocin like inhibitory substances were used as antimicrobial agents that can increase food safety. The curd milk solids can be classified as white cheese that is soft in nature. In this case, fat will be entrapped in the casein that is coagulated. Most of the physical properties of the cheese will be removed from the milk products. We can use proteolytic enzymes to bring about the coagulation process. Most of the water content in the milk can be separated and whey can be removed. The blue cheese and camembert can be classified as cheese that is soft-ripened. We can use Lactobacillus lactis and Streptococcus cremoris species as starter cultures in cheese.

With this biopreservation technique, we can easily safeguard the food products naturally. We can also maintain hygienic quality, and this technique can become a handy tool for most of the food processing industries. We can apply bioprotective cultures in this process. Biopreservation considers foodstuffs that are produced from animals.

The biopreservation systems were used in regulating the spoilage specific organisms. LAB species has minimized most of the food borne diseases that may surface in the food industries. These species can easily halt the production of the harmful microbes. Most of the dairy products can be used in the culturing of strains that may exhibit antagonistic characteristics. This bacterium can withstand acidic conditions. They consist of minimal cytosine and guanine content in the DNA. They can be coagulase-negative and catalase-negative.

We can classify LAB species under the Firmicutes group and in order lactobacillales. We can find lactobacillus species under the Aerococcus genera. Some of the important species that are to be considered are Bifidobacterium, Atopobium, Actinobacteria, Weissella, Vagococcus, Tetragenococcus, Symbiobacterium, Streptococcus, Pediococcus, Oenococcus, Leuconostoc, Lactococcus, Lactobacillus, Enterococcus, Carnobacterium and Alloiococcus species. Most of these species can be classified as probiotics. These probiotics should be used in large amounts to get the positive benefits of the LAB species.

We can categorize LAB species as habitat competitors, and this can be attributed to their competitive exclusion property. There will be release of several antimicrobial metabolites in the product. Most of the microbes can be inhibited, and this can be attributed to the prompt action of the lactic acid bacteria.

Several organisms' acids can be produced by LAB species, and this can be due to the metabolism of sugars. pH values can be lowered because of the generation of organic acids in the medium. Some of the undesirable microbes that can spoil the food products will be inhibited. Hydrogen peroxide can act on the sulfhydryl groups. These groups are often seen in the lipids and membrane proteins. These compounds can easily damage the cell environment but disrupt cell integrity and permeability. The hydrogen peroxide molecules can easily react with the oxygen molecules.

The carbon dioxide molecules can easily reduce the oxygen molecules that are freely available in the medium. Because of shortage of the oxygen molecules, there will be the onset of anaerobic conditions. This will encourage the microbes that can grow in oxygen limited zones. The odor of the milk products can be attributed to the presence of diacetyl molecules. These compounds can also interact with the membranes of bacterial species, and hence the metabolic pathways will be altered.

Most of the bacteriocins released by these microbes can be classified under peptides. These peptides can be produced by both gram-positive bacteria and gram-negative bacteria, and it is our duty to measure the rate at which these microbial species produce bacteriocins.

Most of these peptides can be heat stable and with low molecular weight. The producer cells will be affected on a limited scale, and this can be due to prevailing immunity or resistance mechanisms.

We can also opt for active screening for antimicrobial activity. Paneer is a milk product widely used in developing countries. In preservation of paneer, Lactobacillus plantarum species was used, and it worked as biopreservatives agent and antimicrobial agent.

The bacterial growth was absent at least for a week in the presence of Lactobacillus plantarum. After 7 days, there was a slow bacterial growth, and this is due to the reduced growth of Lactobacillus species. In contrast to this, paneer can be preserved for only one day and without adding any microbial cultures. We can preserve it for six days at freezing temperatures. The shelf life of paneer was increased to 9 days at refrigeration temperatures. The screening for antimicrobial activity results that the Lactobacillus plantarum species can act against Bacillus cereus and Escherichia coli effectively.

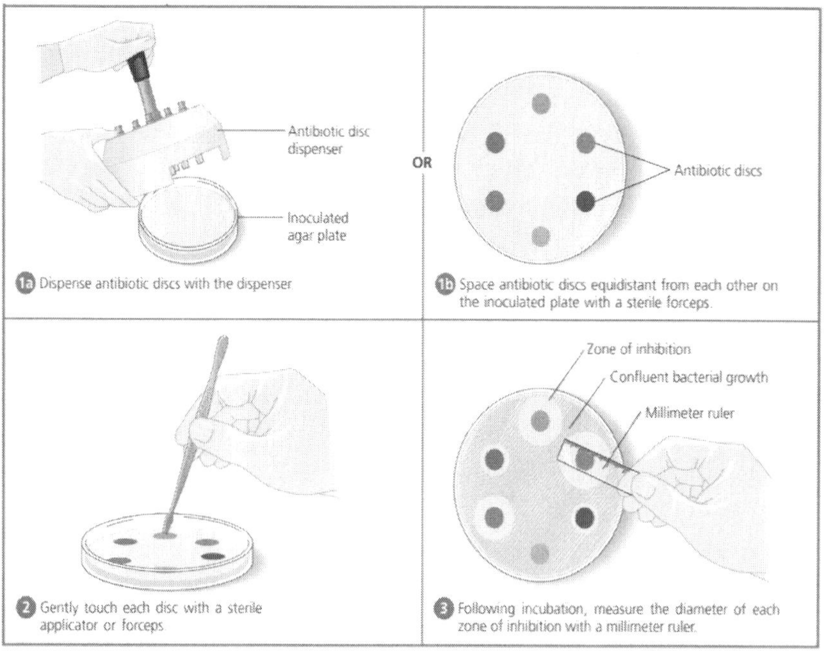

Figure 8.7. Disc diffusion method.

Source: https://i1.wp.com/microbeonline.com/wp-content/uploads/2013/08/Kirby-Bauer-Disc-Diffusion-Method.png?ssl=1.

The antibiotic screening can be done using the disc diffusion method (Figure 8.7). In this method, we can use antimicrobial agents on the bacterial species. We can observe a zone of inhibition that will be about 20.3 mm in diameter. This zone of inhibition is dominant in case of Bacillus cereus. We can get a second zone of inhibition that can be accounted for to 18.5 mm in case of Escherichia coli. We can use water as a solvent. The bacteriocins can be used in enhancing the quality of the dairy products. Bacterial contamination can be of important concern in most of the dairy industries. The main problem is the contamination from the farmed animals, and they can act as reservoir houses for pathogens. Some of the microbes can easily enter the milk products. Some of the species that can spoil the milk products are E. coli, Staphylococcus aureus, Salmonella, and Listeria species. Mostly raw milk can support most of the bacterial species. Pasteurization can easily destroy most of the microbes that are pathogenic in nature. In some cases, after pasteurization, some of the microbes can enter the milk products due to recontamination. Among all these bacterial species, Listeria monocytogenes can damage human health. It will cause flu symptoms followed by gastroenteritis. This condition can be serious and occasionally fatalities may surface. This bacterial species can cause meningitis and bacteremia. In some cases, there will be preterm delivery. Rarely we can notice infant death, neonatal infection, and loss of fetus. Listeriosis can be an important condition that can be caused due to bacterial infestation.

Listeria species can easily contaminate most of the dairy environments, and they can spread through manure. This can occur due to silage that is fermented. Mostly these species can enter via the food supply chain. This pathogen can survive at low temperatures and it can survive in any environment. Most of the soft cheeses can easily support the growth of Listeria species. This species can survive in pasteurized and raw milk. S. aureus can cause mastitis and this condition can be called bovine mastitis. This bacterium can release enterotoxins. Most of these toxins can be thermostable. This bacterial species can cause foodborne illness. Due to this, there will be diarrhea, vomiting, nausea, and abdominal cramps. S. aureus can live in any environment, and mostly we can observe this bacterium in raw milk. Staphylococci can spoil milk by the production of toxins, and they are classified as enterogenic in nature. There will be intoxication due to abnormal temperature fluctuation while storage. E. coli can produce Shiga toxin, and it can produce severe health problems in humans. If the cells are more than 50 in number, then this toxin can spoil the dairy products. E. coli that produce Shiga toxin can be called as STEC serotype. Though the severity of this toxin is high in some cases, the average cases are minimal. This bacterium can cause limited effect on the milk products, and this can be attributed to the best management practices by the farming communities.

Most of these STEC strains can be found in the ruminants and they contaminate the milk through subclinical mastitis. In some cases, fecal routes may be involved. This bacterial species can be present in the milking apparatus that are being used in most of the dairy farms in the developed countries. This species can cause hemolytic uremic syndrome (Figure 8.8) and bloody diarrhea in old people and young children. The main concern with the milk products is the occurrence of salmonella species. This species can cause several health issues in both humans and animals. This species can be observed in the gastrointestinal tract of wild animals and farmed animals. Salmonellosis can occur if this species is not treated properly.

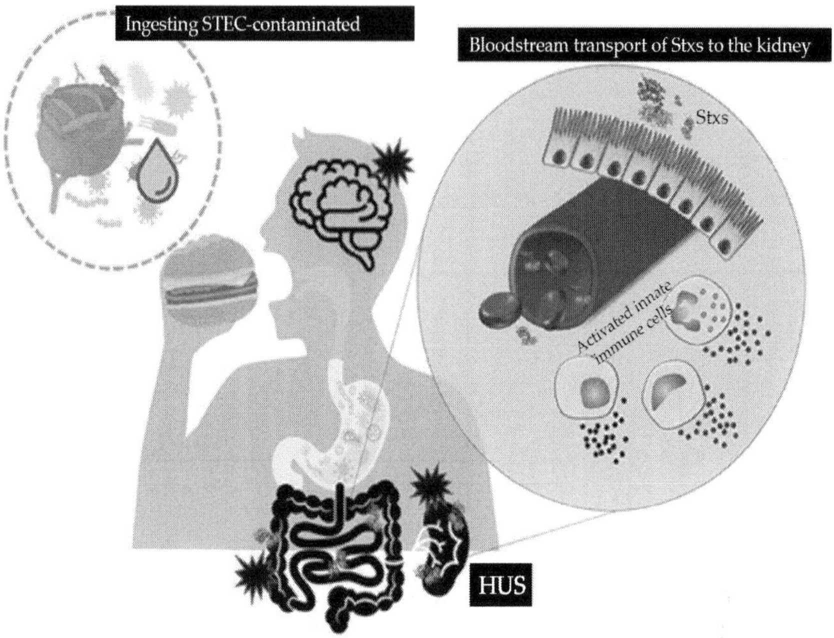

Figure 8.8. Hemolytic uremic syndrome.

Source: https://about-ecoli.com/ecoli_hemolytic_uremic_syndrome.

We need to regulate the growth of salmonella in our food products through appropriate techniques. The occurrence of S. enteritidis has lowered since last decade. This species can infect eggs and meat, and it is considered as one of the common pathogens that spoils the food products. The growth of salmonella can be attributed to the post process contamination and improper pasteurization process. The cheeses that are produced from the pasteurized milk and raw milk will be allowed to age, and this will increase the health risk. There are many factors that can enhance the risk of being infected by salmonella. Most of the population that are susceptible to the infection caused by salmonella are old people, children, pregnant women, and immunodeficient persons. It is always important to reduce the infection chances that might occur among the vulnerable groups. There are several regulatory measures that can be used to lower the incidence of bacterial growth and to reduce the foodborne microbes. There are several hygiene measurements, sanitation and GMP or good manufacturing practices in place, but mostly they cannot avoid the infections.

We can apply bacteriocinogenic strains and bacteriocins in preserving the dairy products. We can apply lactic acid bacteria that can easily produce metabolites and most of the dairy products can be preserved. This will also enhance the quality of the dairy products. We need to use additional hurdles in this case to regulate the bacterial growth. Generally, the bacteriocins can act on related microbes and nonrelated microbes. If these bacteriocins can destroy the related bacterial species, we can call them as bacteriocins with narrow spectrum, and if they act on nonrelated species also, then we can specify them as having a broad spectrum. These bacteriocins can preserve most of the food products. We can classify these peptides into two classes, namely class I and II. We can categorize the bacteriocins that belong to first-class into lantibiotics. These compounds may undergo major modifications in the translational stages. Due to these modifications, there will be addition of thioether amino acids. These thioether amino acids are methyllanthionine and lanthionine. There are several latest bacteriocins that can undergo modifications, and they are not specific to any antibiotics. The bacteriocins that were classified as class II can contain peptides that are unmodified. The class II bacteriocins can be classified into four important groups. They are named as IIa, IIb, IIc and IId. IIa bacteriocins can be called as one peptide pediocin like bacteriocin. IIb bacteriocin can be called as 2 peptide bacteriocins. IIc bacteriocins can be called cyclic bacteriocins. IId class can be called as linear nonpediocin like one peptide bacteriocins. We can add these bacteriocins in combination with the starter cultures in situ. We can also use adjunct cultures in this case. We can use dairy products that are fermented to enhance the safety standards of the food products. We can use pediocin and Nisin in the biopreservation process. We can observe several antimicrobial effects that can be synergistic in nature. With the application of nisin in the dairy products, we can avoid spoilage of several clostridial species. We can observe problems in cheese called late-blowing effects due to contamination. Nisin can kill most of the pathogenic microbes that are associated with Listeria monocytogenes. We can also enhance the effect of bacteriocins by adding sodium chloride and lower pH levels. We can apply lantibiotics in regulating the growth of Listeria species. Lacticin powder can be used to protect several pathogens in cottage cheese and yogurt. We can notice severe antilisterial activity. We can observe the activity and stability of these powders and inactivated bacterial species at different pH levels. We can add pediocin in its dried forms to lower the growth of Listeria species. The bacterial species can be found in cheese sauce systems, cream, and cottage cheese systems.

These bacterial species can easily grow in the neutral food systems and acidic systems that are mild in nature. We can observe the growth of Listeria species in camembert cheese products. The enterocins that are prepared from the bacteriocin producing bacterial species can also drastically reduce Listeria species. Most of the bacteriocins can lose their microbial activity, and this can be attributed to the food matrix. Most of the food components can affect the activity of the bacteriocins in the food systems. We need to consider diverse strategies that can regulate the growth of microbes. Liposomes can be used in the microencapsulation of several bacteriocins, and they have several advantages. In the cottage cheese, we can use Lacticin 3147 bacteriocin that is derived from Lc. lactis DPC 3147 in its spray-dried powder form to combat the growth of Listeria monocytogenes. We can use pediocin bacteriocin that is derived from P. acidilactici PAC1.0 in the form of dry powder to combat the growth of L. monocytogenes in yogurt and cottage cheese. In camembert cheese, we can use bacteriocin called Piscicolin 126 derived from C. piscicola JG 126 in the form of concentrated supernatant to regulate the growth of L. monocytogenes. In the case of goat milk cheese, we can use enterocin CRL35 bacteriocin derived from E. faecium CRL 35 to regulate the Listeria species. This can be used in the form of concentrated supernatant. Nisin can be produced from Lc. lactis CNRZ 150 in the form of starter culture. This can be done to regulate Listeria monocytogenes that contaminate camembert cheese.

Nisin can be produced from Lc. lactis TAB 50 and it can be used as a starter culture. We can use this bacteriocin in case of semi-hard cheese against Listeria monocytogenes. We can use Lacticin 481 & 3147 on cottage cheese and semi-hard cheese against L. monocytogenes. We can add Enterocin AS-48 in case of cheddar and Manchego cheese in the form of starter culture and adjunct culture against Listeria species. Pediocin can be extracted from L. plantarum WHE 92 and Lc. lactis MM 217 in the form of surface sprayed cell suspension and starter culture in case of muenster and cheddar cheese. We can also extract pediocin from L. lactis CL1 in the form of adjunct cultures against L. monocytogenes and S. aureus in semi-hard cheese. Nisin can be produced from L. lactis ESI 515 in the form of adjunct cultures against S. aureus in the semi-hard cheese. The bacteriocinogenic cultures can be used in the form of adjunct cultures and starter cultures in cheese. This will effectively reduce the costs that will be incurred in the biopreservation process. It is worth noting that there is no need for regulatory approval from the food regulatory agencies. We can also use tolerant and resistant cultures of nisin producing strains. These characteristics can be considered as desirable properties. This can substitute the commercial forms of nisin. We can use Lc. lactis species in the form of mixed starter cultures that can produce nisin. We can isolate nisin producing strains in the production of cheese, and they can be purified from raw milk cheese and raw milk. There are several enterococcal bacteriocins that can be classified under class II pediocins, and they can provide strong antimicrobial activity against Listeria species. It is essential to safeguard foods according to the strains that are being used in the biopreservation process. We can use purified bacteriocins in preserving food products that can be susceptible to the pathogenic bacteria. These compounds can be used in the place of antibiotics. We need to ensure that the commensal microbiota of humans should be safe.

The microbiota of the gut can be important as it will improve immune function, metabolism, and digestion. There are variations in the function and diversity of the micro-ecosystem of the gut. The disorders in the microbiota of the gut and intestine can bring about several infectious diseases, obesity, insulin resistance, immune diseases that can be inflammatory and bowel disorders. Most of these gut related problems can be attributed to the development of Clostridium difficile species. We can easily tackle dysbiosis and pathogenic species using probiotics in our diet. Probiotics are nothing but the microbes that are in their living state. These are to be supplied in huge quantities so that there will be an appropriate effect on the gut and intestines.

Most of the intestinal pathogens can be regulated on usage of probiotics in appropriate quantities. Antimicrobial action against the pathogenic microbes can be considered as an essential trait of the probiotics that we can use. Due to the inclusion of probiotics, we can modulate the immune system, and there will be competition among the dominant bacterial species for nutrients. The main issue with the use of probiotics is that most of the gastrointestinal diseases and their relationship with probiotics is yet to be established. We can use Bifidobacterium and Lactobacillus species in the form of probiotics in our diet. In the Gastrointestinal tract, these bacteriocins can make the bacterial strains go to their dormant stage. These peptides are highly active against most of the pathogens. These peptides can be called antimicrobial peptides that can work as signaling peptides. This can be done via quorum sensing. There will be cross talk between several microbial communities. These can send signals in the activation of the immune systems. In the murine infection model, the action of Lacticin and nisin was studied. Most of the bacteriocins can be classified as quorum sensing molecules. Some of them were classified as autoinducing peptides in the intestines. The nisin molecules can be produced as signal molecules, and in most of the cases, they will be secreted from bacterial species. Due to these signaling molecules, most of the genes will be transcribed.

We can use chemical preservatives like sodium nitrite and sodium nitrate in the preservation of the dairy products. These compounds can cause severe health effects, and they are to be addressed immediately to compensate damage already caused. These compounds may also result in the production of the carcinogenic compounds due to their cross-reactivity. Most of these pathogens can reduce the shelf life of these milk products, and these chemical preservatives will work well to balance the pathogens. Most of the literature suggested that bacteriocins can serve well in the biopreservation processes. These compounds can be heterogeneous in nature. We need to evaluate carefully prior to their selection as they are species specific. Most of the spoilage specific bacteria can be controlled by the addition of these bacteriocins. The cost of bacteriocins is quite high, and this will not justify the costs that are to be invested in storing the food products. The production of bacteriocins dates to 1925 when Gratia discovered "proncipev." This compound controlled the growth of another harmful strain of E. coli. In 1946, Gratia with Fedricq used "concine" for the same purpose. Later, these compounds were named as bacteriocins. These compounds can exhibit maximum affinity against the proteins produced by other bacterial species.

These compounds can exhibit narrow range activity against several pathogenic bacteria. Most of the gram-positive bacteria may not contain specific receptors that can easily attach with these bacteriocins. Lactococcus lactis can produce Nisin that can be classified as class I/ type AN antibiotic. Staphylococcus epidermidis can produce epidermidin and Staphylococcus gallinarum can produce Gallidermin. L. lactis can be used in the preparation of Lacticin 481. Actagardin can be produced from Actinoplanes sp. The bacteriocins called as duramycin, Ancovenin, Cinnamycin and Mersacidin can be classified under class I type B lantibiotics. Duramycin can be produced from S. cinnamomeus and Ancovenin can be produced from Streptomyces spp. Cinnamycin can be produced from Streptomyces cinnamoneus species. L. lactis can produce lactococcin MMFII and Divercin V41 can produce Carnobacterium divergens. Enterocin A can be produced from Enterococcus faciem species. Leucocin A-UAL 187, Sakacin Pm Sakacin A and pediocin PA-1/AcH can be class IIA bacteriocins. Platarcin JK, Plantaracin EF, Plantaracin S, Plantaracin A, Lactacin F, Lactoccin M, Lactococcin G can be classified as class II B bacteriocins. Class II B group can constitute two complex peptide groups. These compounds can exhibit minimal activity. Most of these compounds can act on the membranes of bacterial species. These compounds contain amphiphilic properties, and this is the reason why the bacteriocins can damage the cell membranes. Mostly, there will be formation of pores. The Lactic acid bacteria can generate many peptides that can create pores in the cell membrane. These compounds can be easily conducted, stable and size can affect the physicochemical properties of bacteriocins.

These compounds can produce poration complexes. Most of the inorganic phosphates will be leaked, and this will cause an imbalance of ions. There will be stabilizing interaction between the cationic residues and phospholipids that are components of the peptides. Most of the hydrophobic regions will be present in the external leaflet of the cell membranes. Most of the bacteriocins will get aggregated on the surface of the membranes. This bacteriocin complex can spread all over the membranes, and there will be the formation of transient membranes. The proton motive force can be dissipated.

8.4. BACTERIOCINS

Most of the bacteriocins will target sensitive cells only, and the cytoplasmic membranes will be primarily targeted. The translocation of these molecules into the bacterial cells is still to be understood. Lysis is another condition that bacterial cells will endure in response to action against these bacteriocins. Most of the gram-positive cells will be targeted by these bacteriocins. Some of the bacterial cells can be producer strains, resistant strains, and sensitive strains also. Nisin can be used as a surfactant, and this can be attributed to the cationic nature. This will increase the permeability of the bacterial cells so that the UV light can easily get into them and destroy the components of the cell. The Nisin compound can easily detect the phospholipids present in the cell membrane. There is no need for a specific substrate like platform so that nisin can get attached. Lactose Trypsin and Lactococcin A need certain receptors so that they can easily bind onto the membrane lipids. Most of the cells that are susceptible will be targeted. IIA bacteriocins can detect the components of the cell membrane, and there will be enhanced permeability. This can be due to the presence of amino acids with dominating N terminal regions. Plantaracin JK and EF (Figure 8.9) can be categorized as class IIB bacteriocins. Their activity will be based on the peptide types, and they can easily interact with the receptors present on the cells. Most of them can produce pores and hence there will be dissipation.

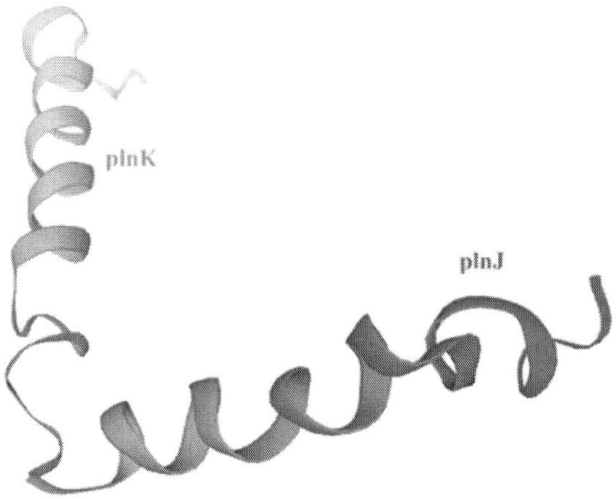

Figure 8.9. Plantaracin JK.

Source: https://microbialcellfactories.biomedcentral.com/articles/10.1186/s12934-020-01365-6/figures/4.

There will be a certain rate of natural resistance that can be attributed to the alteration of the surface phospholipids. Some of the molecular components can be in contact with the bacteriocins. In some cases, there will be increased resistance, and this can be due to the formation of mutant strains. The alterations in the cell envelope can also bring about altered fatty acid composition in the membranes. The genes that encode the permeases can be interrupted. These permeases can be called phosphotransferase system permeases or PTS permeases. Resistance to bacteriocins was prominent in Bacillus cereus, Clostridium botulinum, Staphylococcus aureus, and L. monocytogenes. The main reason behind this resistance can be due to changed receptor structures and the action of proteases. In some cases, bacteriocins will be destroyed due to enzymes produced from the bacterial species. We can also observe change in the shape of the membranes. We can add these bacteriocins through three modes. We can apply these compounds into food products via in situ production and this can be through adjunct cultures and primary cultures. We can use purified or semipurified bacteriocins in the production of food products. We can incorporate the ingredients that can be fermented previously from the bacterial strains. We can use thermal processing methods in the food processing sector, and this can affect the bioactivity of the bacteriocins that are being applied in the food products. We need to consider fat content and pH levels before applying these compounds that are derived from the bacterial species. There are many ingredients that can be used by the microbes to lower the activity of the bacteriocins. Using lactocin 705, sodium chloride, ascorbic acid, alginate, sodium nitrite and sodium lactate can reduce the effectiveness of the bacteriocins. Out of all these compounds, lactocin 705 can inhibit the pathogens. Addition of sodium chloride can also reduce the bactericidal activity of these bacteriocins. Ethylenediamine tetraacetate can be used to disrupt gram negative bacteria, and this should be food grade. Addition of EDTA can increase the susceptibility of bacterial species to bacteriocins. EDTA can bind with the magnesium ions present in the lipopolysaccharides of the bacterial cells, and hence the membrane can be easily disrupted.

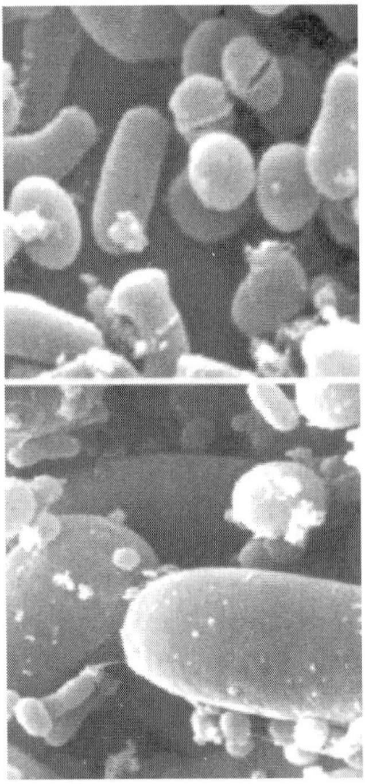

Figure 8.10. Staphylococcus equorum.

Source: https://www.sciencedirect.com/topics/agricultural-and-biological-sciences/staphylococcus-equorum.

Staphylococcus equorum WS 2733 (Figure 8.10) can produce bacteriocins, and they can be inoculated into soft cheese. With the application of this compound, the Listeria species can be inhibited even at low contamination levels. Lactococcus lactis IFPL359 can produce bacteriocins and can work as transconjugants. It can be used in the form of starter culture in the cheese produced from goat's milk. On application of this culture, we can speed up the ripening process of cheese. Lactococcus lactis CNRZ481 can be classified as bacteriocin producing species that can be used as adjunct culture. We can add this culture into cheddar cheese. There will be release of Lacticin 481 from this bacterial species, and this will induce cell lysis in bacteria. Enterococcus fecalis can be used in the form of adjunct culture. We can use this culture in nonfat hard cheese and skimmed milk.

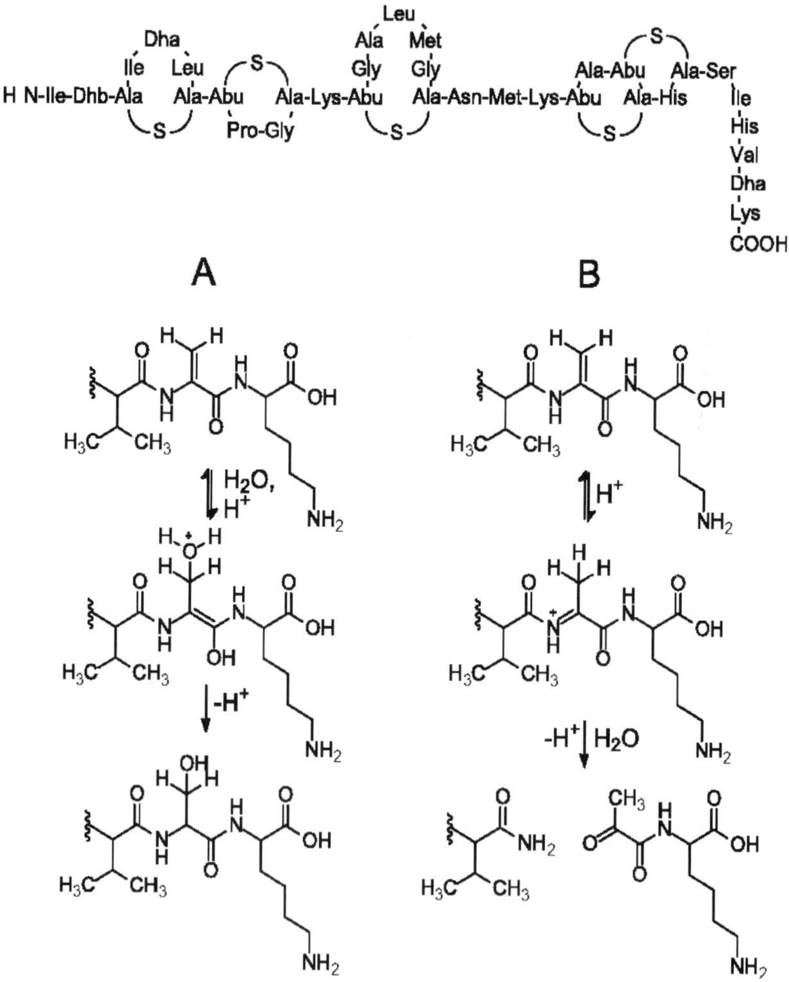

Figure 8.11. Nisin Z.

Source: https://www.sciencedirect.com/science/article/abs/pii/S0308814611000811.

Nisin Z (Figure 8.11) can be produced from Lactococcus lactis WB, and this can be applied to whole fat milk or skim milk. This addition has reduced the growth of Listeria monocytogenes. Nisin A and Z and Lacticin 481 can be produced from Lactococcus lactis species in the cottage cheese. This can be used against Listeria species but with limited positive results. Nisin can be used in Galotyri PDO cheese and this will prolong the shelf life of cheese that is stored at freezing temperature for a week. Nisin A can be used to preserve milk pudding. It can regulate the spore forming bacterial species. It will also extend the shelf life of the milk pudding. We can use minimal heat treatments to enhance aroma and flavor, and the food safety can be ensured. Nisin can be added to Minas Frescal cheese and this will effectively reduce the growth of S. aureus species in both whey and curd. The effect of bacteriocin can be limited. Reuterin can be isolated from Lactobacillus reuteri species and it can be used to store the UHT skim milk. We can combine both nisin and reuterin to minimize the growth of S. aureus and L. monocytogenes. This can be done at 4°C to 8°C. Leucocin can be produced from Leuconostoc mesenteroides to preserve the UHT-whole fat milk. We can combine glycine and leucocin to inhibit the Listeria species. Variacin can be isolated from Kocuria varians and this was used in the preservation of most of the dairy food models.

8.5. NISIN

Nisin can be active only if the pH values are lower than 7, and it is obvious that its activity is effectively reduced in alkaline conditions. There are many problems with the use of nisin in dairy products. Nisin react with the fats that constitute most of the food products. It is to be noted that the interaction between the fat molecules and nisin is not well established. Cream when heated exhibited high rates of inhibition of Bacillus cereus species at minimal concentrations of nisin. These nisin molecules can easily regulate the spoilage bacteria and participate in extending the shelf life of the products. This is true in case of pudding made up of milk with higher fat content. The homogenization process that is being used in the sterilization of milk can also lower the effect of nisin on the Listeria species. The sterilization techniques can effectively damage the bacteriocins that are otherwise used to store the food products for long periods. In these practices, all materials are produced from lactic acid bacteria or related species. If we use lactic acid bacteria in the fermentation of the foods, then the use of nisin in their preservation will be difficult. We can opt for other bacteriocins that can bring about more effect in the case of preserved foods. We need to consider resistance against proteolytic enzymes, heat tolerance and stability parameters of the food products. We need to investigate the antibacterial activity also. Most of the pediocins can be classified under class IIa bacteriocins. These compounds can be produced from Pediococcus species. Pediocin can be more effective than nisin in controlling S. aureus and L. monocytogenes species. These microbial cultures can be classified as food-borne pathogens. We can also observe other pathogenic bacterial species like E. coli and Pseudomonas species.

These bacterial species can be classified as gram negative bacteria. The pediocins can be active at various pH ranges and they are resistant to heat treatments. These compounds can remain active in the freezing and heating conditions. These characteristics can be beneficial in preserving the dairy products.

These compounds can easily reduce the growth of Listeria monocytogenes species in cheese sauce, cream, and cottage cheese. This activity is possible only if the Listeria species exist in limited numbers. We can observe this action of pediocins in diverse pH ranges and temperatures. We can produce pediocin (food grade) from cheese whey that is already fermented. We can use semi purified pediocins in regulating the growth of S. aureus species. These compounds can enhance the shelf life of raw milk obtained from buffalo.

Lacticin are important bacteriocins that can act on the pathogenic bacterial strains. These compounds can be produced from L. lactis. We can isolate Lacticin 481 and Lacticin 3147 compounds from this species. We can also isolate Lacticin 3147 from Irish kefir grains. These grains are being used in the preparation of buttermilk. Lacticin 3147 can be classified as a 2-component lantibiotic. It needs structural proteins in order to obtain biological activity. These bacteriocins can act against several microbes.

This bacteriocin can also reduce the growth of the food spoilage bacteria and pathogenic bacteria per se. The activity of this compound can be in line with the action of lactic acid bacteria. We can prepare Lacticin 3147 in the form of powder preparations.

These powder preparations can regulate the growth of Bacillus and Listeria species that might contaminate the milk for infants, cottage cheese and natural yogurt. The cost of this compound can be an important factor in its application in preserving the dairy products as it seems to be costlier. Another single peptide lantibiotic that can be discussed here is Lacticin 481.

This compound can act on some microbes and this activity can be classified as medium spectrum. This compound can attack Clostridium tyrobutyricum and LAB species. Along with these bacterial species, we can also expect antagonistic property against Listeria species. Lacticin 481 can be used in its impure form. It can exhibit limited bacteriostatic activity in case of milk products that are stored in the refrigerators. Application of Lacticin 481 in its impure or semi pure form reduced the growth of Listeria species for at least 7 days. It is worth noting that this bacteriocin itself might not be effective against all the pathogens that are present in the dairy products.

Enterocins can be considered as an important class of bacteriocins that can be used in milk-based products. These compounds can be isolated from Enterococcus species. They have a broader inhibitory spectrum. This bacterial species can produce different types of bacteriocins. The bacteriocins that are produced from enterococcus species can be cause for concern, and this can be due to diverse opinions on its GRAS status. Enterococci can cause several health problems in humans. Most of these species are resistant species and virulent ones. Some of these species can easily express antibiotic resistance genes. The enterococci species that are food-grade can exhibit limited virulence, at least at the clinical level. These species can easily exchange their genetic information among other species in the microbiome. It has made it mandatory to use purified forms of enterocins in the food consumption, and this can be attributed to the rising concerns about their virulence. Enterococcus fecalis can produce enterocin AS-48, and this can be classified under class IIc of bacteriocins that are cyclic in nature. These bacteriocins can easily act against Clostridium and Bacillus species. This compound can exhibit a high range of stability at diverse pH levels. They are also stable at high temperatures. This is the reason why the compounds can be used in the food products without any alterations in their structures. We can control the growth of Staphylococci species in the milk that is already skimmed. We can also expect a bactericidal effect while using this bacteriocin in the food products. The concentration of bacteriocins can be directly proportional with the antibacterial activity of these bacteriocins. We may need a high concentration of these compounds to eliminate the staphylococcus species from the dairy products. Enterocin KP can be used against Listeria species in full fat milks, half fat and skimmed milk. This compound can exhibit enhanced antagonistic effects on the Listeria species, and we can also expect bactericidal effects at a broad spectrum. If the amount of Listeria monocytogenes concentration gets increased along with the fat content of the milk, we can observe reduced activity of this bacteriocin. The purified enterocins can exhibit enhanced inhibitory effects against most of the microbes that can easily grow in the milk products. After 1 day of application, the effect of enterocin has reduced, and this can be attributed to several factors that are in action. Most of these compounds can destroy Clostridium and Listeria species and the recontamination may not be prevented on application of these compounds. We can use enterocins and reuterin bacteriocins that can initiate combined effects on the food products that are to be stored. This effect can last for at least one day. It is to be noted that most of the inhibition rates can be related to the dosage that is being served to the food products. This condition can be observed in case of fresh cheese. If we add high doses to the cheeses, we can expect high rates of inhibition. This inhibition can be active for at least 3 days.

There are other miscellaneous bacteriocins that can inhibit the microbial growth in the milk products. Lactococcin BZ can be isolated from L. lactis species. This compound can act against both Gram-positive and Gram-negative bacteria. This compound exhibited firm activity against Listeria species. We can also use this compound in its partially purified forms. The antilisterial activity of this compound can be active at 20°C also. This activity will remain for at least 25 days and it can be extended. The fat content of the milk products might not be affected by the action of this bacteriocin.

Aureocin A70 is another essential bacteriocin that can be categorized under class II bacteriocins. This compound can be produced from S. aureus. We can easily isolate this species from commercial milk. This bacteriocin can induce a bactericidal effect on gram-positive bacteria. This can be observed in the UHT-milk also. Listeria species can remain controlled for at least 7 days due to the addition of this bacteriocin. We can also eliminate the pathogens completely.

We can also use genetic engineering to improve the efficacy of the bacteriocins. Ent35-MccV is one such bacteriocin that can be classified as chimerical bacteriocin. This is produced by genetic engineering techniques. Through this technology, we can combine microcin V that is active against E. coli species and CRL 35 that is active against Listeria species. This genetically engineered bacteriocin is active against both E. coli and Listeria species.

Though the application of most of the bacteriocins preserved dairy products, their usage in bulk quantities is prohibited or in this case, limited. Bacteriocins may not provide complete protection from the pathogenic species per se. The cost factor is another limitation that limits their use in the dairy industry. Due to limited approvals from the agencies like EFSA and FDA, bacteriocins utilization remained restricted to pediocin and nisin. These compounds can be commercially available in most of the areas. The European food authorities imposed restriction on the use of Enterococcus species. This species can be present in most of the dairy foods and in some cases, we can use it in the form of starter culture. This species has become a member of the microbiota of,most cheeses, especially artisanal cheese.

There are many health stimulating factors in the milk products. This is the reason why milk products are in high demand. Through diverse methods, milk will provide benefits to humans. Lactobacillus can be used in promoting human health. These bacteria can act as immune modulator. It can act as an anticancer agent. These bacteria can also exhibit calcium-binding nature, and it can act as an antihypertensive agent in most of the cases. L. lactis subsp. lactis can be used as a starter, and we call it as a mesophilic starter.

This bacterium can be used in the preparation of buttermilk, butter, and cheese. Lc. Lactis subsp. lactis biovar diacetylactis can be used in the preparation of buttermilk, lactic butter sour cream, edam, and gouda. Lc. lactis subsp. cremoris can be used as a mesophilic starter in the preparation of buttermilk, butter, and cheese products. Streptococcus thermophilus species can be used as thermophilic starter. We can use this species in cheese products and yogurt. We can also use this bacterial species in hard high cook cheese and in some cases even in semihard cheeses.

Lactobacillus acidophilus species can be used in the form of probiotic adjunct culture. We can use it in the preparation of yogurt and cheese. Lactobacillus delbrueckii subsp. bulgaricus can be used as a thermophilic starter in the preparation of cheese and yogurt. We can use this species in the preparation of hard and semihard cheese. Lactobacillus delbrueckii subsp. lactis can be used in the preparation of high cooked cheese and fermented milks also. Lactobacillus casei can be used in the form of thermophilic starter in fermented milks. We can use this species in the preparation of cheese. Lactobacillus plantarum can be used in probiotic milk products. We can use this bacterial species as an adjunct culture in the ripening of the cheese. Lactobacillus rhamnosus can be used in the form of adjunct culture in the preparation of cheese.

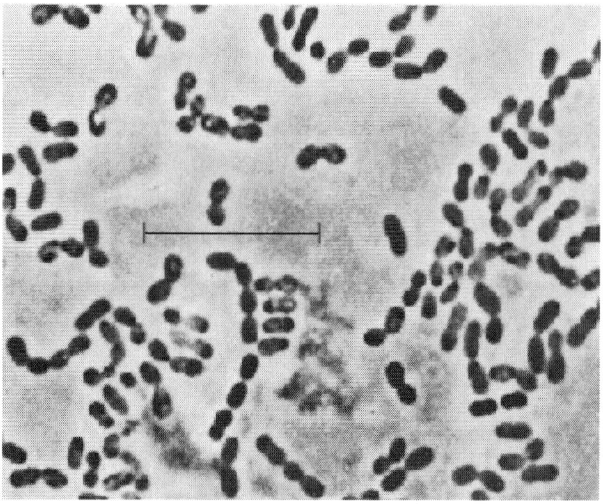

Figure 8.12. Leuconostoc mesenteroides subsp. Cremoris.

Source: http://cooknkohlmesenteroides.pbworks.com/w/page/16454569/classification%20.

We can use Leuconostoc species in the preservation and processing of the dairy products. Leuconostoc mesenteroides subsp. cremoris (Figure 8.12) can be used in the form of mesophilic cultures. This species can be used in the preparation of sour cream, lactic butter, fresh cheese, Gouda, and edam products. If we increase the load of lactic acid bacteria in the milk, then we can obtain lowered lactose concentrations. This implies lowered content of harmful bacteria. Due to this, there will be negligible bloating and diarrhea in humans. The health of the gut will be improved because of rejuvenation of the normal microflora on ingestion of fermented products. Several fungal infections in women and bacterial vaginosis can be prevented on usage of fermented products as foods. We can gain a tendency to easily digest lactose content in the milk products and can be used as a source of energy. Eating of fermented foods can also ease digestion. Some of these bacterial species can serve as antihypertensive agents. The milk products will be cleaved into short amino acid chains. We can easily detect casein peptides once these products are subjected to fermentation. Most of these peptides can exhibit essential functional properties that can be beneficial to humans. These products can also help us in relieving pain and hence can be used as natural analgesic. Some of these compounds can enhance our immunity levels. These compounds can also prevent growth of the pathogenic species.

Due to the calcium-binding properties present in the fermented foods, we can also expect bone strengthening and building properties. There will be an increase in acidity due to fermentation and hence we can expect sharpness in taste. Due to increase in acidity, pathogenic species will be killed.

Apart from the preservation using microbes, we can also use physical methods that can enhance the shelf life of the milk products. High pressure processing or HHP can be considered as one of the important techniques that can be used in the preservation process. This process was demonstrated by Hite. This process makes use of Le Chatelier's principle and Isostatic principle. According to Le Chatelier's principle, the pressure parameter will encourage the structural reactions, and this will vary with decreasing volume. In case of isostatic principle, the pressure will be distributed, and it will be passed onto all the portions of the food material no matter what size and shape they are in. In this process, the food materials will be subjected to high pressures. We can start with at least 100 MPa in this process. We can use oil mixture or water as a transmitting medium to impart pressure on the food products. We can use this process in the sterilization of milk, cheese, and other beverages. We can easily deactivate microbes, and this can be attributed to damage that can be caused due to the application of high pressures. The cell wall and membranes can be damaged with these pressures.

We can use pulsed electric fields also in the sterilization of the food products. In this process, we can place the food products in between the electrodes. There will be application of short electric pulses. The high voltage of 80 kV will be supplied through these electrodes. These high voltages can be produced using pulse generators. Due to the application of these high voltages, there will be electroporation that can be observed on the cytoplasmic membrane. We can use continuous systems and static systems in this case. We can use this technique for milk products. In case of milk products, we can use 50 kV. This process can activate enzymes and enhance shelf life in the milk.

Cold plasma is another technique that can be used in the preservation of dairy products. Mostly we refer to plasma as the fourth state of matter. This state can be classified as neutral electrically. This can be produced by application of gas and electricity. Some of the products of plasma are nitrogen species, OH ions or radicals, nitric oxide, NO_2, NO, and N_2. This plasma can be produced in atmospheric temperature, and then we can call it cold plasma. This can be produced through diverse methods. We can use

Dielectric barrier discharge method, gliding arc discharge, corona discharge, and atmospheric plasma jet discharge. In the generation of the electric field, we can use noble gases like helium, neon, and argon along with nitrogen and air. In the dairy industry, we can easily use the Dielectric barrier discharge method. This method can be used in destroying several pathogens in slices, cheese, raw milk, and cheese. On application of this strong electric field, we can observe radical bombardment in the cell. There will be denaturation of enzymes and proteins. The cell membrane will suffer with lipid oxidation. We can also notice damage of both RNA and DNA. There will be formation of lesions in the internal cell environment. The cold plasma may not be used in the decontamination and preservation process, and this can be attributed to limited penetration capability of cold plasma (Figure 8.13).

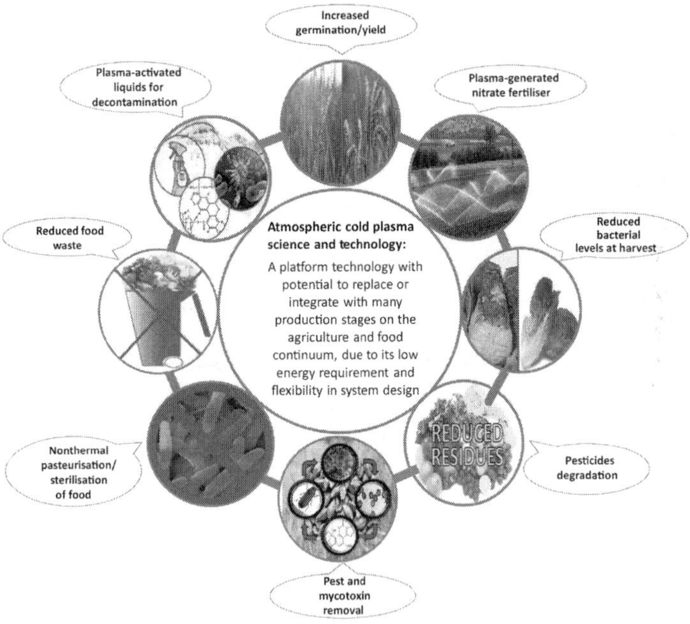

Figure 8.13. Cold plasma.

Source: https://www.cell.com/trends/biotechnology/fulltext/S0167-7799(17)30 299-8.

8.6. ULTRASONICATION

Ultrasonication can be considered as an essential technique in destroying several microbes that can spoil the food products. In this process, we can use sound waves that can reach a frequency greater than 16 kHz. These sound waves will be passed on through gases, solids, and liquids. After passing these sound waves, we can observe small bubbles in the food products, and this can be called cavitation. It is important to know that the bubble size increases. In some cases, these bubbles can get imploded as a result of the application of high intensity waves. Whilst implosion, we can observe increased pressure and temperatures. We can opt for the thermosonication process. This process combines ultrasonication and heat. This process can result in the removal of most of the food products. We can also combine ultrasonication with pressure. This process can be called manosonication in most of the food producing sectors. We can combine sonication, pressure, and heat to decontaminate the foodstuff. This combined process can be called manothermosonication. We can also observe the inactivation of cells, and this can be due to disruption of the cell wall structure. Due to induced cavitation, the functioning of the cell will be impaired. This can be observed if the food is subjected to ultrasonicators based processes. Irradiation is another crucial practice that can be used in the decontamination process. In this process, food will be subjected to ionizing radiation. This radiation can be electron beams, ultraviolet light, X-rays, and gamma rays. It is important to keep the irradiation levels below 10kGy. Most of the microbes will be killed on application of UV light. The microbial count in the food products will be reduced on application of UV light. Most of the food products will be protected from harmful microbes. The UV radiation can be classified or categorized into UV-A, UV-B and UV-C. The UV-A type will have wavelength ranging between 315 and 400 nm. UV-B radiation can possess wavelengths ranging from 280 to 315 nm. UV-C type has a wavelength from 200 to 280 nm. Out of all these UV ration types, UV-C can be considered as an essential radiation type that can be used in the preservation of the food products. The transparency and composition of the food products can influence which UV light penetrates. This UV radiation can damage DNA in the bacterial cells, and there will be an occurrence of mutation. The shelf life of cheese will increase due to irradiation, and this can be due to inactivation of microbes. Ohmic heating can be used in the preservation and decontamination process. This technique has been used since the nineteenth century. The apparatus in which we can use ohmic heating can be called a joule heater. In this apparatus, we can observe the migration of ions, and this will occur dominantly in the liquids. Due to the bombardment of ions, we can expect high collision rates. Due to these collision rates, there will be formation of resistance and heat. The shelf life of the milk products can be increased if they are subjected to ohmic heating. The shelf life of the milk products can be enhanced with the addition of endolysins, bacteriophages, natamycin, lactoferrin, lysozyme, and bacteriocins.

8.7. HERBS

The herbs and spices can be used in the preservation and treatment of milk products. We can use rhizomes and roots of turmeric in this process. These plants can produce vitamins, essential oils, antioxidants and phytosterols. All these substances can be used in the regulation of the microbes. Some of the essential oils can be produced from the plants and can be used in lowering the growth rate of several microbes (Carmen et al., 2015). These substances can extend shelf life of ghee and other milk products. There are many chemical preservatives that can be used to protect the milk products. We can use chemical agents that can exhibit antimicrobial activity. Benzoate, sodium salt and benzoic acid can be used in the preservation of milk-based products. We can use inorganic salts, sugar, and salt in the decontamination and preservation process. We can use specialized packaging material and packaging systems in the preservation process. These packaging materials can be used in both transportation and distribution processes. This packaging material can protect the food products easily. We can use smart packaging, intelligent packaging, edible coating, active packaging, modified atmosphere packaging and vacuum packaging.

In the case of modified atmosphere packaging, there will be alteration of the native gaseous environment. This can be important in the packing of the foods. Some of the dominant gases that can be used in this packaging are oxygen, nitrogen, and carbon dioxide. We can actively and passively modify this atmosphere that can be used in packaging. This technique can control the growth of the microbes and also halts the deterioration of the foodstuff. The packed dairy products like yogurt, jalebi and cheese can be preserved with the aid of this technique. Active packaging is an important technique in the regulation of the microbes. We can add active components into the materials and lower the microbial growth rates. These components can work as active ingredients. We can call these components oxygen scavengers. These ingredients can act as antioxidants and like substances that can absorb and emit carbon dioxide. We can also add zinc oxide, Ag nano-composite and other essential oils to extend the shelf life of the milk and cheese. Edible coatings can be used in the decontamination and preservation of the dairy products. The edible packaging or coating of the food materials can be biodegradable in nature. This method is almost the same as the traditional packaging material. Some of the materials that can be used in the edible coating process are resins, lipids, protein and polysaccharides. We need to opt for secondary packaging also. This packaging can be important in hygienic management of the food products. We can use polylactic acid, galactomannan and chitosan as the materials that can be used in the secondary packaging process.

Compounds produced from cardamom, cumin, cinnamon, cassia bark, coriander, and black pepper can be used in the preservation of the dairy products. All these compounds released have medical properties. There are several spices that were used in the preservation of the food products. It was reported that small cardamom, garlic, red chili, ginger, and turmeric exhibit antimicrobial effects, and these herbs are considered as safe for human consumption. Some of the dairy products like yogurt, sour cream, and cheese can be produced by a fermentation process assisted by lactic acid bacteria. This bacterium was used as starter cultures in most of the dairy industries. We can use Lactobacillus helveticus, and Lactococcus species as starter cultures. The probiotics can effectively enhance the nutritional quality of the foods. Most of the dairy materials will be subjected to the bacterial inoculation to enhance nutritional content. Most of the lactic acid bacterial species can work as probiotics. We can use Bifidobacterium and lactobacillus species along with probiotic strains to enhance the food products. The preservation of milk products can be pursued by several methodologies that are diverse. Most of the probiotics produced by inoculating these bacterial species can improve human health. Most of the prebiotics can work well and are comparable with the probiotics. Probiotics are nothing but living yeast and bacteria, and these organisms can extend the shelf life of the food products. We can use probiotics in order to manage ailments like colon cancer and allergic reactions. We can use these probiotics along with nuts, seeds, olive oil, cooked quinoa, avocado, yogurt, fruits, eggs, and protein powder.

Most of the food industries are using chemical preservatives to save the food products. Most of these chemicals can enhance the shelf life of the milk products. Some chemicals that can be used in this process can induce hazardous health effects. They can also disrupt the dietary importance of the food products. We can use sulfur dioxide and nitrates in this process. Antimicrobial peptides can be used in the preservation of the food products. These compounds can also enhance the aroma of the food products. These chemicals can be toxic to humans, and it is obvious that we need to use natural preservatives. These days people are using recombinant synthesis, enzymatic synthesis, chemical synthesis and peptide synthesis in the preservation of dairy products. We can categorize the bacterial species that can induce undesirable changes. They are pathogenic bacteria that can cause diseases in humans and animals, and the second type is spoilage bacteria that disrupts taste, odor, and flavor. Most of the spoilage bacteria can be found in common places, and the exposure levels of the dairy products have to be considered in the quantification of spoilage. Most of the symptoms will surface on ingestion of these products. The primary principle of the entire preservation process lies with killing of the microbes and enzyme control. It is also important to use all the physical methods needed to decontaminate the milk products along with the biopreservation methods. In the raw side domain of the entire preservation cycle, there will be 'receiving side' and 'storage side.' In the processing domain of the food industry, there will be pasteurization, sterilization, and dehydration stages. In the packaging and distribution domain, there will be filling machines, delivery mechanisms, and refrigerators. It is to be noted that the use of chemicals in the preservation process is both costly and time-consuming business. The antioxidants, antimicrobial agents, inorganic acids, and organic acids can be costly processes. Most of these chemicals were restricted and prohibited due to their harmful effect on humans.

Chapter 9

Biopreservation of Meat Products

CONTENTS

9.1. INTRODUCTION

The shelf life and product quality of the meat products and meat itself can be based on the biochemical and microbiological acumen of the carcass that we use in the processing. It is worth noting that the raw materials should be kept in near pristine condition. We need to check the quality of meat right from the slaughterhouses. The slaughtering animals can play a crucial role in providing us with excellent protein sources. There are many qualities in the meat products that can get disturbed or disrupted due to external factors and internal factors. Once these qualities are disturbed, we may not obtain the actual product. We need to ensure proper handling of carcass and hygienic quality should be maintained all along. Some of the slaughter techniques that are in vogue in the rural areas are still far from sophistication. Because of improper handling of meat, there will be lessened bleeding in the animals. This can result in the storage of blood in the muscles. Because of increased blood content in the muscles, overall shelf life will be reduced that can be attributed to the degree of acidity. Improper covering of the meat products can also contaminate them with microbes and dust. The dirty equipment, tools, and pollution can be considered as hot zones that can spread a lot of spoilage specific agents that can effectively reduce the shelf life of the meat products. If the evisceration process is not done properly, there will be rupture of tripes and stomach, and this can lead to spread of bowel microbes to all surfaces of the carcass. In some cases, carcass may get split over the ground and this will increase the contamination risk. The unclean containers, bars, hooks, and clothes can contaminate the meat products during transit. The unhygienic conditions may prevail due to rough handling at the processing plants, shops, and meat markets. Due to unhygienic practices at various stages, the microbial load in the meat products will be contaminated, and there will be the onset of health hazards. Some of the microbes can produce toxins, and their production is negligible in the fresh meat. Rural slaughtering is a common source of microbial contamination. Most of the slaughter operations that are to be done on the carcass must be at hanging posture. The slaughtering of large animals can invite several challenges pertaining to microbial contamination. At the hanging position, we need to allow the animal carcass to bleed. We can also incise their hind feet at this posture. We can remove the skin of the legs and belly at this posture itself. It is better to use ropes and wooden structures in the essential stages like splitting, evisceration, and skinning. These structures can be used in deboning and cutting also. In some of the rural areas, slaughter slabs are being used for small animals and slaughtering at the field is prohibited. We must always hold the animal carcass suspended from a metal frame or wooden frame, and this will keep it far from the floor. In most of the slaughter slabs, we can observe platforms that are made up of concrete. It is expected to be at least 50 cm from the ground, and it is essential that it dips so that the draining will be easily possible. We need to install a roof so that the fresh meat may not get affected by external weather conditions. They need to be fenced too. We need to keep accessible outlines for draining the effluents. Scalding vat can be used if pigs are to be slaughtered. At the slaughterhouse, we need to install a Lairage, floor, chiller, tripe room, meat cutting room, processing room, blood disposal facility, solid waste disposal, skin processing, hides processing, disposal of effluent, light and power supply, meat market and water supply. Once the slaughter carcasses are prepared, we need to shift it to the chilling room. We need to use the chilled meat for processing. In some countries, there are limited refrigeration facilities. Prior to cutting, we need to investigate the signs that can be due to microbial contamination and spoilage. When there is unchilled meat, it is important to prepare dried meat. Heat treatment can be an appropriate method in case of unchilled meat. In the case of hot meat, we need to use trimmings and cuts immediately after the slaughtering process is done. Most of the meat surfaces can be treated with sodium sorbate, ascorbic acid, tartaric acid, citric acid, lactic acid, and acetic acid. We can add these chemicals in the form of sprays and dipping solutions.

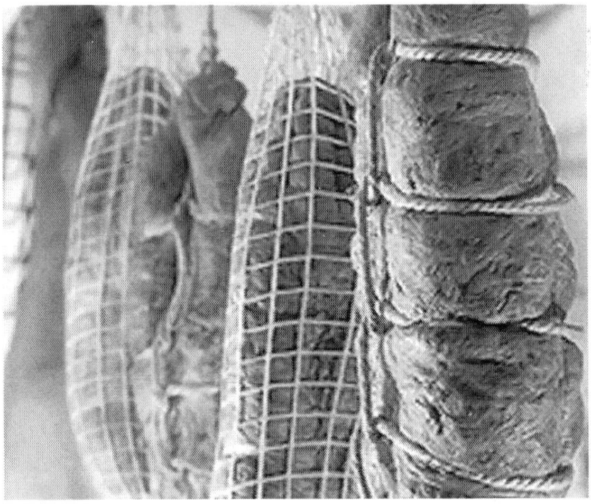

Figure 9.1. Preservation of meat.

Source: https://www.dartagnan.com/preserved-meat-methods.html.

9.2. TRADITIONAL METHODS

The oldest method of meat preservation (Figure 9.1) involves drying meat in natural temperatures, humidity, and air movement, including direct sun rays. It involves gradually dehydrating chunks of meat cut to a specified uniform shape, allowing full batches of meat to be dried evenly and simultaneously. The ideal circumstances for meat drying are warm, dry air with a low humidity of around 30% and modest temperature variations between day and night.

Figure 9.2. Meat drying.

Source: https://travel.mongabay.com/laos/images/laos_2047.html

Meat drying (Figure 9.2) can, however, be done successfully under less favorable conditions if basic hygienic and technological guidelines are followed. The drying process' intensity and duration are influenced by air temperature, humidity, and circulation. High temperatures, low humidity, and good air circulation will speed up the drying process. Evaporation of water from the meat's peripheral zone to the surrounding air, as well as continuous migration of water from the deeper flesh layers to the peripheral zone, reduce the meat's moisture content. During the first day of drying, there is a significant amount of water evaporation from the meat, which thereafter gradually diminishes. Weight losses of up to 60%–70%, comparable to the quantity of water evaporated, can be noted after drying the beef for three or four days. As a result, by monitoring the weight of a batch during drying, moisture losses can be monitored. Through shrinkage of the muscle and connective tissue, continuous evaporation and weight losses during drying induce changes in the shape of the meat. The meat portions get smaller, thinner, and wrinkled to some extent. In addition, the consistency shifts from soft to firm to hard. In addition to these physical changes, several biological reactions have a significant impact on the product's organoleptic properties. In underdeveloped nations, meat for drying is frequently taken from unchilled carcasses, and rapid ripening processes occur during the first stage of drying as the meat temperature remains relatively high. As a result, the distinct flavor of dried meat differs significantly from that of fresh meat. The characteristic flavor of dried meat comes from a slight oxidation of the animal lipids (Figure 9.3).

Figure 9.3. Animal lipids.

Source: https://en.wikipedia.org/wiki/Animal_fat#/media/File:Reuzel.jpg.

Only lean meat should be dried in general. The quality of the finished product is harmed by visible fatty cells sticking to muscular tissue. Rancidity develops quickly in dry meat under processing and storage conditions, causing flavor degradation. Meat from camelids, sheep, goats, and venison (e.g., antelopes, deer) are also used in the production of dry meat. Meat from a medium-aged animal that is in good shape but not obese is the best for drying. Meat from animals in poor nutritional condition can also be dried, however the increased amount of connective tissue increases hardness. Raw meat for the production of dry meat must be thoroughly inspected for undesired changes such as discoloration, hemorrhagic patches, off-odors, parasite manifestations, and so on. It is necessary to remove such flaws. To acquire meat suited for drying, carcasses must be carefully chopped. Beef corpses are more difficult to manage in rural environments than sheep, goats, or game carcasses due to their size. Beef carcasses must be chopped and deboned right after slaughter if chilling facilities are not available.

Because meat is always served with a small amount of salt, the raw material might be salted before drying. This method not only improves the taste of the product, but it is also more technologically and hygienically advantageous. This is done with pure common salt, either dry or dissolved in water. A 14-percent salt solution is ideal when drying meat that has been chopped into strips or flat sections. The purpose of dipping the meat in the salt solution is to prevent microbial growth on the surface of the meat. As a result, salting must be done within five hours of slaughter, because after that time, tremendous microbiological development develops, which is unaffected by salt treatment. Second, the result protects against insects during the drying process. Various insects, particularly domestic flies, are attracted to the freshly cut meat surfaces, which feed on the moisture released by muscle fibers. These insects contaminate the meat heavily and may even lay their eggs in it. After being soaked in the salt solution, meat is no longer an appealing setting for insects. The high salt concentration on the surface of the meat keeps them at bay.

During the drying process, a thin layer of crystalline salt forms on the surface of the flesh. Because the salt crystals are hygroscopic, they absorb some of the water excreted by the meat, keeping the flesh surfaces dry. One reason for the preservability of presalted and dried meat is that dry meat surfaces hinder the growth of germs and molds.

To avoid contamination and maintain a long shelf life for the dried product, the beef strips must be handled under rigorous hygiene conditions prior to drying. However, if particular pieces are accidentally contaminated, further processing can only be done with extreme caution. After rinsing the contaminated pieces of meat in clean water, a special bucket with salt solution should be available to soak them in. It should be noted, however, that the original quality of contaminated items cannot be restored. As a result, such pieces should always be dried individually and used as soon as possible in the preparation of meals, rather than being stored for a long time. The traditional method of suspending meat for drying by hanging strips from tree branches, wire, or rope is not suggested since the meat remains in contact with these supporting devices or may come into contact with one another, causing the meat to not dry correctly in these contact regions. As a result, the preferred way is to suspend the meat strips individually from one end, ensuring unrestricted air circulation throughout the whole length of the pieces and quick and uniform drying through proper drying facility layout. Contact between meat pieces while drying must be avoided at all costs, since these areas will remain moist and humid for an extended period of time, providing an ideal environment for spoilage, germs, and flies.

Meat carcasses are handled and kept in most underdeveloped nations without or with relatively modest refrigeration. Meat murdered throughout the night and morning is sold and devoured the same day, thanks to a fast-turnover system. Because microorganisms cannot grow to the point of rotting the meat and rendering it inedible in the short time between slaughter and consumption, this well-proven traditional approach works properly. This approach is ideal for rural villages and small towns in many developing countries, especially because local cuisine does not require mature and soft meat. Although huge contamination happens often during slaughtering and meat handling, this high turnover of meat usually does not result in meat deterioration. Fortunately, hazardous microorganisms rarely proliferate to the point where they endanger the health of customers, as long as the time between slaughter and cooking is kept short and the meat is cooked properly and at sufficiently high temperatures. Fresh meat is usually not packaged at all or is wrapped in paper, leaves, or other natural materials in traditional meat handling. Meat, which is historically kept by drying, is sometimes packaged in linen bags, baskets, or ceramics to make storage and transportation easier and to protect it from dirt, insects, and other contaminants. However, as population density rises, this traditional technique is becoming obsolete in several emerging countries, as more time is required between slaughter and final consumption. Meat is routinely stored, transported, processed, and distributed by a shop or supermarket, all of which takes a significant amount of time. Certain preservation methods must be used to keep fresh meat safe for such a long period of time. The apparent remedy is refrigeration, but it is costly and thus commonly unavailable in impoverished countries. As a result, energy-saving storage solutions are especially important in developing countries. Proper packaging plays a vital role in both techniques of extending the shelf-life of meat and meat products, whether employing refrigeration technology or energy-saving approaches.

Microbial development in meat and meat products will not occur under the circumstances described above. This does not, however, imply that the items will remain stable for an indefinite amount of time. Chemical or physical deterioration, as well as rancidity and discoloration, will shorten their shelf life. In this case, the use of appropriate packing materials that decrease physical and chemical influences on the product or entirely protect it will benefit the product quality. There is a chance you'll be exposed to harmful influences.

9.3. OXYGEN LEVELS

The oxygen level of the air is approximately 20%. When oxygen comes into contact with meat and meat products for an extended period of time, it turns red into grey or green and causes oxidation and rancidity of fats, resulting in unwanted off-flavors. The oxygen permeability of foils used for food packaging varies. The better the preservation of product quality, the lower the oxygen permeability of the packing material. Oxygen-proof packing films combined with vacuum packaging of the goods will provide the optimum protection. This ensures that virtually no oxygen remains in the package and that no oxygen from the air enters the goods.

Because light provides the energy for these processes, prolonged exposure of meat and meat products to daylight or artificial light increases oxidation and rancidity. Transparent packing films do not provide any protection against light. As a result, colorful or opaque films should be used for objects that are exposed to a lot of light. Light is completely impervious to films laminated with aluminum foil. When maintained in the dark or under moderate illumination, products in transparent packing film are well protected. If fresh items with a high moisture content, such as meat, fresh sausages, cooked ham, and so on, are not packed, they will lose a significant amount of weight and quality due to evaporation during storage. As a result, the packaging material must be adequately vapor-proof. This criterion is met by most plastic films used in food packaging. Contamination of meat can be avoided to some extent during slaughtering, carcass dressing, meat cutting, and/or processing. Packaging alone will not prevent microorganisms from multiplying in meat and meat products. Secondary contamination of these meals, such as by contact with dust, dirty surfaces, or hands, can be avoided entirely with adequate packing, particularly with plastic sheets that are impermeable to secondary contamination agents.

Meat is a perishable product due to its high-water content, dense nutritional matrix, and almost neutral pH. These circumstances are ideal for the growth and multiplication of spoilage microorganisms and diseases found in food. Consumer needs nowadays, because of increased education and awareness, include minimal processing items that keep natural flavor, color, and texture, as well as high quality, convenience, innovation, regularity, and a "freshness"-like look with assurance of health elements. To satisfy these desires, new nonthermal approaches are being developed. Meat and meat products are high in protein, have a good balance of vital amino acids, vitamins, minerals, and have a high biological value, making them an ideal home for spoilage microorganisms and food borne diseases to develop and multiply. Consumers nowadays expect minimally processed foods that preserve natural qualities and have a "freshness" appearance. Irradiation is a nonthermal method for killing spoilage bacteria and foodborne diseases while preserving the natural flavor of food. Meat that has been irradiated has a longer shelf life. During irradiation, however, the qualitative qualities of meat change.

Most studies concluded that at medium dosages, irradiation has no discernible influence on the sensory qualities and nutritional value of meat. The quality of the product is affected by radiation dose, dosage rate, temperature, oxygen concentration, and other factors. The color of muscle food is influenced by the packaging environment. At lesser doses, no appreciable protein loss occurs; however, at greater doses, some amino acids are lost, and numerous by-products such as ammonia, hydrogen, carbon dioxide, and amides are generated, resulting in an off-flavor and odor. Radiolytic breakdown of Sulphur amino acids produces Sulphur volatiles that induce irradiation off-flavor. Meat and meat products contain a large amount of water. Water produces a variety of reactive ions and free radicals when exposed to radiation. Due to the limited chances of obtaining other dietary components, the effect of these reactive compounds reduces in the frozen state due to the synthesis of original compounds created by the interaction of ions and free radicals.

Figure 9.4. Beef.

Source: https://www.britannica.com/topic/beef.

Beef (Figure 9.4) and its derivatives have high protein content and a wide range of amino acids, making them popular with consumers both at home and abroad. Beef is a significant source of protein in the daily diet, as it contains a variety of amino acids necessary for human metabolism. Furthermore, the peptide produced by the bovine belt can help to enhance human metabolism and keep it healthy. Beef is also high in iron, tin, zinc, copper, manganese, and other trace minerals. These trace elements play a significant role in human metabolism and antioxidant enzyme production. Beef's fatty acids and fat-soluble vitamins are essential components of a healthy human diet. China produced 6.89 million tons of beef in 2018, ranking third in the world and accounting for 10% of global production. The amount of cold fresh meat in domestic fresh meat production and sales has gradually increased in recent years. As a result, prolonging the shelf life of chilled meat is critical to the industry's success. Microorganisms can easily taint and degrade fresh meat during cold storage. The main causes of frozen beef deterioration are oxidation and microbiological contamination.

9.4. ESSENTIAL OILS

Cinnamaldehyde, cinnamyl ester, salicylaldehyde, and other compounds are the main elements of cinnamon essential oil. Because of its broad-spectrum antibacterial and good antioxidant characteristics, cinnamon essential oil has been frequently used in preservation studies in recent years. Thyme essential oil is derived from the plant's stems and leaves. When thyme blossoms, it emits a powerful odor. Thyme has the ability to suppress germs as well as scavenge free radicals. Thyme essential oil not only has a powerful inhibitory impact on common bacteria in the body, but it also removes free radicals from food and prevents oxidation. Ginger is a vital life spice that is commonly used in traditional Chinese medicine clinical treatment. Clove has recently been discovered to be a natural food preservative as a medicinal herb and spice. Many studies have shown that clove volatile oil is a natural antioxidant and antibacterial agent with good antibacterial and antioxidant effects. Clove essential oil has a distinct smell, is volatile, insoluble in water, and light, oxygen, and temperature-sensitive. Clove essential oil's use in food preservation is severely limited due to these reasons. Microencapsulation of core materials can successfully tackle this problem by improving its stability in food processing and masking its poor flavor. Cumin is a spice that is commonly used to season meat, but it can also be used in the processing and preparation of a variety of meals. It is a natural medication and food source. Cell shrinkage, local deformation, cell wall and membrane rupture, and cell system loss were all observed in bacteria treated with natural essential oil, according to the findings. Flavonoids, volatile oil, alkaloids, and other compounds make up the majority of Zanthoxylum bungeanum. These compounds are abundant in plants and hinder the growth of spoilage and harmful microorganisms in food. Zanthoxylum bungeanum (Figure 9.5) is expected to become a significant source of natural food preservatives.

Figure 9.5. Zanthoxylum bungeanum.

Source: https://www.cnseed.org/zanthoxylum-bungeanum-seed.html.

Meat is the meat of animals that is consumed as sustenance. Fresh meat refers to meat from recently processed animals, as well as vacuum-packed meat or meat packed in controlled-atmospheric gases that has not been treated other than chilling to assure preservation. Meat is a good environment for the growth and propagation of meat spoilage microorganisms and common foodborne diseases due to its complex nutritional makeup. As a result, appropriate preservation procedures must be used to ensure its safety and quality. Although additional methods of preservation are explored to minimize other deteriorative changes such as color and oxidative changes, the technologies used in meat preservation are primarily concerned with limiting microbial decomposition.

The shelf life and keeping quality of meat are influenced by a number of interconnected elements, including holding temperature, ambient oxygen, endogenous enzymes, moisture, light, and, most critically, microorganisms. All these elements, individually or in combination, can cause undesirable alterations in meat color, odor, texture, and flavor. Although meat can deteriorate in the absence of microorganisms due to processes such as proteolysis, lipolysis, and oxidation, microbial growth is by far the most essential component in maintaining fresh meat quality. Traditionally, meat preservation methods have been classified into three broad categories based on temperature, moisture, and, more directly, inhibitory processes such as bactericidal and bacteriostatic, such as ionizing radiation, packing, and so on, but a single method of preservation may employ many antimicrobial principles. Each control step can be thought of as a "hurdle" against microbial proliferation, and hurdle technology is a collection of techniques that can be used to accomplish specific microbiological and organoleptic quality goals.

Natural ventilation chilling, in which refrigerant is circulated through cooling tubes, is slow and uncontrollable, whereas forced-convection air chilling, which uses fans to circulate the air, is far more efficient. Due to lesser evaporation from the surface, rapid carcass freezing promotes product yield, while rapid drying of the carcass surface helps to decrease bacterial growth. On the other hand, ultra-rapid freezing of prerigor beef can cause cold-shortening and toughening. Spray-chilling can increase surface myoglobin oxygenation without increasing metmyoglobin, resulting in a brighter look and no weight loss. Fast freezing creates minute intracellular ice crystals, which reduce leak when thawed. The rate of freezing is affected not only by the weight of the meat and its thermal qualities such as specific heat and thermal conductivity, but also by the temperature of the refrigerating environment, the type of refrigeration, and, in the case of smaller portions of meat, the type of the wrapping material used. Because of the significant temperature differential between the cryogen and the meat product, as well as the rapid rate of surface heat transfer caused by the cryogen boiling, cryogenic freezing enables faster freezing times than standard air freezing. Cryogenic freezing does not necessitate the use of mechanical refrigeration equipment; all that is required is a cryogen tank and appropriate spray equipment. However, the cryogenic procedure may produce some distortion of the product's structure, which could have an impact on its economic usefulness. Furthermore, because cryogenic liquid is relatively expensive, its commercial usefulness may be limited.

During prolonged storage at cool temperatures, carbon dioxide and ozone have been used to inhibit the growth of surface microorganisms on cattle carcasses. Even though ozone leaves no poisonous residues in meat, its usage in a manufacturing environment can be hazardous to workers. It also has a faster rate of lipid oxidation and is more effective against airborne microorganisms than those on meat. Different species produce organic acids and alcohols through anaerobic fermentation of dietary substrates, and they can help preserve food by suppressing other organisms that are present and could deteriorate or render it hazardous. Lactic acid, for example, is a common inhibitory agent used in the preservation of fresh meat; however, other organic acids have also been implicated in discoloration and the generation of unpleasant odors. Salts like sodium lactate have been employed in the meat business for their potential to enhance flavor, extend shelf life, and improve product microbiological safety. Lactates have antibacterial properties due to their capacity to reduce water activity and the lactate ion's direct inhibitory impact.

Organic preservatives such as essential oils, chitosan, nisin, and lysozyme have been studied as alternatives to chemical preservatives to create "green label" products. Natural or managed microflora, of which lactic acid bacteria (LAB) and their antimicrobial compounds such as lactic acid and bacteriocins have been widely studied, can improve storage life and boost safety. Bacteriocins are antibacterial proteins with a diverse range of activity, mode of action, molecular weight, genetic origin, and biochemical features. Preservative characteristics of several spices and essential oils have been employed to extend the storage life of meat products. Cloves contain eugenol, and mustard seeds contain allyl isothiocyanate. Chitosan is a polysaccharide that has antifungal and antibacterial effects. Its efficacy, particularly when used in combination with other antimicrobial drugs, demands additional research. Nisin, the only commercial bacteriocin, has been used to disinfect intentionally contaminated raw pig parts and as an antilisterial agent in minced raw buffalo meat when combined with 2% sodium chloride.

Bacteriocins, spices, and essential oils are examples of bioactive surface coatings on packaging materials that may have activity based on migration or evaporation into headspace. Antimicrobial activity was found against the majority of the indicator microorganisms in four polyethylene films varying in ethylene-vinyl acetate and erucamide content and coated with three different bacteriocins, with antimicrobial agent dispersion and film roughness associated to packing activity. The introduction of antimicrobials such as bacteriocins and plant extracts into active packaging and their relationship with biodegradable packaging such as alginate, zein, or synthetic polyvinyl alcohol (PVA) to reduce waste and be environmentally friendly is one of the most promising fields. Bioactive edible coatings combine an antibacterial ingredient with an edible coating that is applied to food by dipping or spraying. Edible polysaccharide, protein, and lipid coatings can improve the quality of fresh, frozen, and processed meat and poultry products by delaying moisture loss, minimizing lipid oxidation and discoloration, improving product appearance, and acting as a carrier for food additives, among other things.

Oxidative and microbial/enzymatic deterioration are not prevented by freezing. As a result, chemical preservation methods can be used in conjunction with refrigeration to improve product stability and quality while preserving freshness and nutritional value. To avoid the formation of rancidity, off-flavor, and other undesirable effects, a thorough understanding of lipid oxidation and its suppression is required. Autolysis is indeed a word used to describe a sequence of postmortem chemical reactions in animal tissues after death induced by the actions of enzymes such as lipolytic, amylolytic, and proteolytic, which are responsible for the metabolic process during the life of animals and for the destruction of lipids, carbohydrates, and proteins after death. Fat deterioration, also known as lipolysis or oxidation, is caused by lipolytic enzymes, whereas glycogen to lactic acid conversion is caused by amylolytic enzymes. These alterations take place during the initial stages of storage. Protein deterioration is caused by proteolytic enzymes that convert proteins to amino acids and then to amino nitrogen or nonprotein nitrogen, resulting in an increase in the soluble nitrogen products of the meat.

Meat is the human's preferred option of animal protein, and meat consumption is constantly rising around the world. In comparison to the

1960 s, yearly per capita consumption climbed by 2.6 times in 2000 and would increase by 3.7 times by 2030. The rich nutrition matrix meat, on the other hand, is susceptible to numerous types of deterioration depending on handling and storage conditions. Every year, large amounts of meat and meat products deteriorate at the consumer, merchant, and restaurant levels, resulting in significant economic and environmental losses. Due to the breakdown of vital substances such as fat, protein, and carbs, meat rotting results in the development of bad flavors, off aromas, and often slime formation, making the food unfit for human consumption. As the world's population grows and the food supply becomes more globalized, meat spoilage must be controlled to extend its shelf life and maintain its nutritional content, texture, and flavor. Meat and meat products can be improved in quality and shelf life by using proper handling, pretreatment, and preservation processes.

Low temperature storage and chemical treatments are the most used in the business today for controlling enzymatic, oxidative, and microbiological deterioration. It is crucial to keep meat below 4°C immediately after slaughtering, as well as during transportation and storage, for sanitation, safety, shelf life, appearance, and eating quality. At a lower temperature, however, microbiological and enzymatic deterioration can be stopped or limited. However, freezing will not prevent oxidative deterioration. For limiting deterioration of meat and meat products, a combination of chemical additives such as Tertiary Butyl Hydroxy Quinine (TBHQ) and ascorbic acid can be most efficient. Antimicrobial and antioxidative properties are shared by these chemical additions. TBHQ inhibits lipid oxidation, whereas ascorbic acid inhibits microbiological deterioration. The most practicable combination to minimize spoiling and increase the shelf-life of meat and meat products is the inclusion of preservative chemicals like TBHQ combined with ascorbic acid and storage at refrigerated temperatures under darkness. To improve the shelf-life of meat, further research is needed to understand the effect of animal age, animal type, stress level before and during the slaughtering process, initial microbial load, type and behavior of bacteria, and their interactions.

Refrigeration is the most extensively used method of meat preservation for short-term storage because it slows or limits the pace of spoiling at temperatures below the optimum range, inhibiting microbiological development, enzymatic and chemical reactions. Fresh meat is kept refrigerated at a temperature of 2 to 5°C. Meat cleanliness, safety, shelf life, appearance, and nutritional quality all benefit from chilling. Carcasses are initially chilled in coolers to remove body heat before being transferred to holding coolers. It is critical to provide enough space between carcasses to allow for optimum air circulation. Meat chilling begins with the cooling of the animal corpse and continues throughout the entire supply chain, including holding, cutting, transportation, retail, display, and even at the customer's home before final usage. To avoid excessive shrinkage owing to moisture loss, the relative humidity is usually controlled at 90%. The species of origin, initial microbial load, packing, temperature, and humidity conditions during storage all affect the refrigerated storage life of meat. Pork and poultry have a higher microbial load than other meats. Regardless of the species of origin, extreme caution should be exercised when handling meat to prevent further microbial infection. Refrigerated temperatures encourage the growth of psychrophilic microbes, which leads to the deterioration of meat over time. Fresh meat can be kept refrigerated for up to 5-7 days if maintained at a temperature of 4°C. Ultra-rapid freezing of prerigor beef might cause cold-shortening and toughening. It's important to remember that processed meat should be kept refrigerated until it's time to eat it. When compared to fresh meat, well-preserved meat has a longer shelf life.

The freezing rate has an impact on the quality of frozen meat. Huge ice crystals occur during slow freezing, which can inflict physical harm to muscle tissue and give it a deformed look in the frozen condition. When meat is frozen quickly, countless little ice crystals grow uniformly all through tissue. With a drop in temperature, the freezing rate increases; at −20°C, nearly 98% of water freezes, and at 65°C, total crystal formation occurs. As a result, there is no concern with muscle fiber loss or deformed appearance in meat tissue. Because water freezes within the muscle fiber itself, drip losses during thawing are minimal. Fast frozen meat has a lot of little ice crystals on the surface, which gives it a nice bright tint when contrasted to slow frozen meat. Microbial development stops around −12°C, and cellular metabolism in mammalian tissues is completely inhibited below −18°C. Enzymatic processes, oxidative rancidity, and ice crystallization, however, will continue to play a role in spoiling. Approximately 60% of the viable microbial population dies after freezing, whereas the remaining microbes gradually rise while frozen storage.

The key curing components are sodium chloride, sodium nitrate, sodium nitrite, and sugar. Dry cure, pickle cure, injectable cure, direct treatment, and other methods of curing are used. Meat preservation by excessive salting is a tradition. Insufficiently enough quantities, sodium chloride has a long history of usage in food preservation. Because refrigeration facilities were not accessible in the past, it was used. Later, curing with ordinary salt and sodium nitrate produced items that were noticeably better. By increasing osmotic pressure and lowering water activity in the microenvironment, sodium chloride limits microbial development. Concentrations as low as 2% have been shown to inhibit certain bacteria. Many food spoilage yeasts, such as Debaryomyces hansenii (Figure 9.6), Yarrowia lipolytica (Figure 9.7), Kloeckera apiculate (Figure 9.8), Kluyveromyces marxianus (Figure 9.9), Pichia anomala (Figure 9.10), Pichia membranaefaciens (Figure 9.11), Saccharomyces cerevisiae (Figure 9.12), Zygosaccharomyces bailii (Figure 9.13), and Zygosaccharomyces.

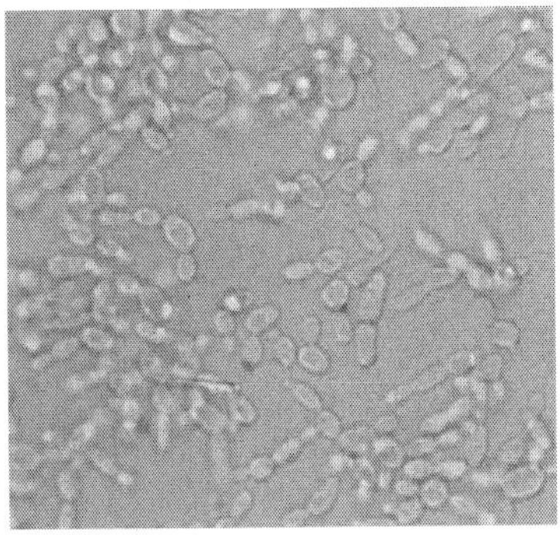

Figure 9.6. Debaryomyces hansenii.

Source: https://upload.wikimedia.org/wikipedia/commons/1/11/Candida_wet_mount.jpg.

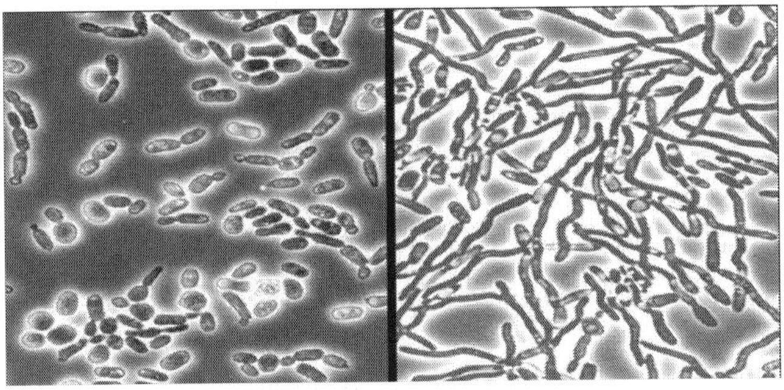

Figure 9.7. Yarrowia lipolytica.

Source: https://www.researchgate.net/publication/287293867_Dimorphism_and_Pathogenicity_of_the_Opportunistic_Ascomycota_Candida_albicans/figures?lo=1.

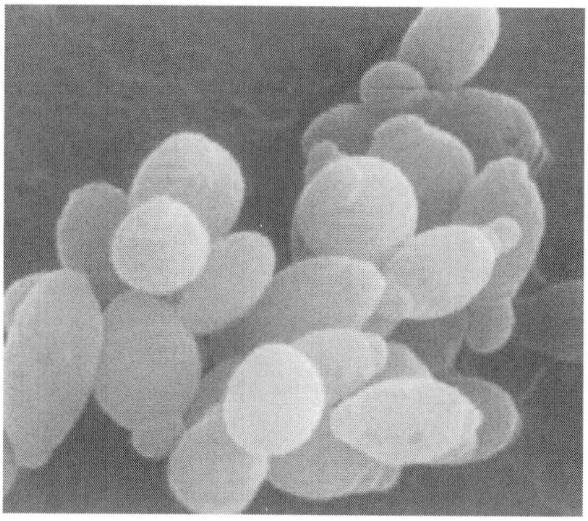

Figure 9.8. Kloeckera apiculate.

Source: https://www.semanticscholar.org/paper/Behavior-of-Kloeckera-apiculata-Flocculent-Strain-Sosa-Nadra/905cba39a8cbc68d17ef658d59271db95f3a759a/figure/4.

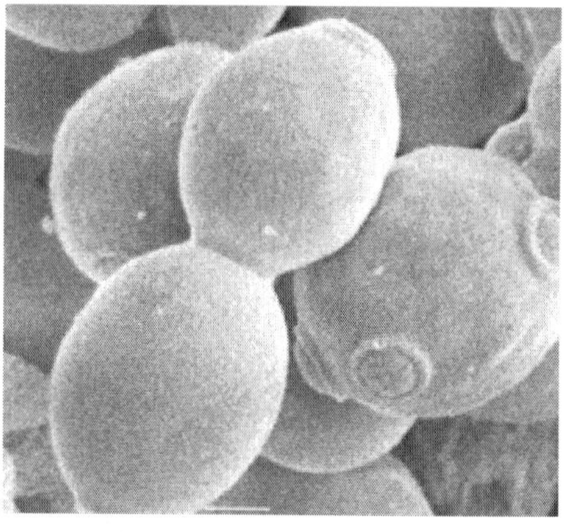

Figure 9.9. Kluyveromyces marxianus.

Source: https://www.sciencedirect.com/topics/biochemistry-genetics-and-molecular-biology/kluyveromyces-marxianus.

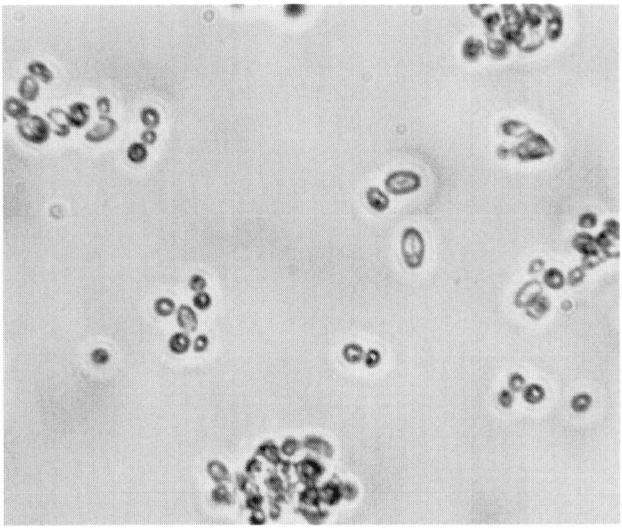

Figure 9.10. Pichia anomala.

Source: https://wineserver.ucdavis.edu/industry-info/enology/wine-microbiology/yeast-mold/pichia-anomala.

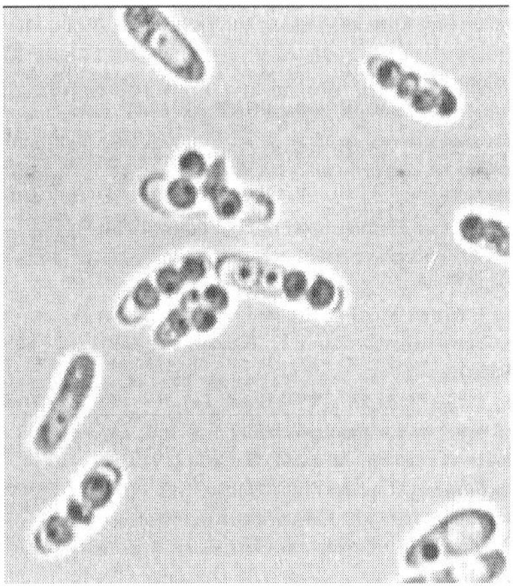

Figure 9.11. Pichia membranaefaciens.

Source: https://mycocosm.jgi.doe.gov/Picme2/Picme2.home.html.

Figure 9.12. Saccharomyces cerevisiae.

Source: https://en.wikipedia.org/wiki/Saccharomyces_cerevisiae.

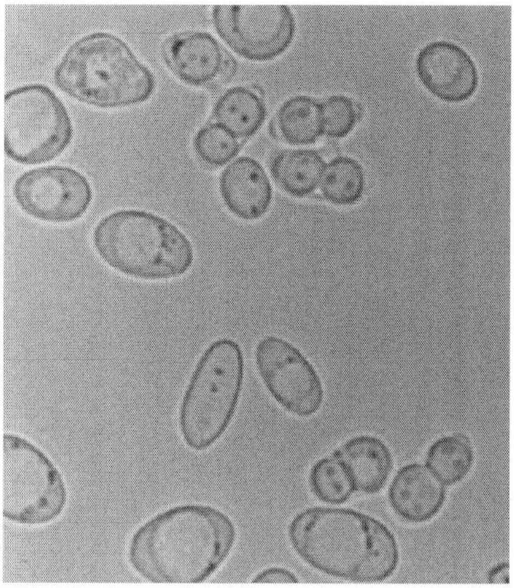

Figure 9.13. Zygosaccharomyces bailii.

Source: https://www.sciencedirect.com/topics/immunology-and-microbiology/zygosaccharomyces-bailii.

Some Bacillus and Micrococcus bacteria, on the other hand, have demonstrated the ability to tolerate high salt concentrations. Sugars have the ability to bind moisture and reduce the amount of water in foods. Dextrose, sucrose, brown sugar, corn syrup, lactose, honey, molasses, maltodextrins, and starches are commonly used as sugars or carbs in dry meat processing to improve flavor, minimize salt harshness, and lower water activity. Sugars are usually regarded as safe in Canada and the United States. Nitrites are always in the form of salts in the meat preservation sector, such as sodium nitrite or potassium nitrite. Nitrites help to keep the color of red meat stable, as well as the flavor of cured meat and the prevention of rancidity.

Figure 9.14. Oregano.

Source: https://www.thespruce.com/herb-profile-oregano-1761786.

Thyme and oregano (Figure 9.14), among other herbs, are widely used in Mediterranean cooking, particularly in meat recipes. Despite the fact that the essential oils of these two plants have powerful antibacterial and antioxidant characteristics, their use as natural meat preservation is limited due to their hydrophobicity, sensitivity to external variables, and interactions with food components. Essential oils can also have a negative impact on the organoleptic qualities of meat. Incorporating essential oils into nanometric delivery devices could be one way to get through these roadblocks. Essential oils are stabilized, protected, and released in a controlled manner when they are nano-sized. By minimizing undesired interactions with dietary components, the essential oils' bioavailability is improved, and their potential negative impact on the organoleptic qualities of meat products is reduced. Numerous researches have proven the antibacterial and antioxidant properties of nanoencapsulated essential oils, and some of them suggest that essential oils were effective in food models such as beef burgers, pâté, and rainbow trout in this form. Incorporating nanoforms of essential oils in packaging methods, on the other hand, is a more promising technique to incorporate them into foods.

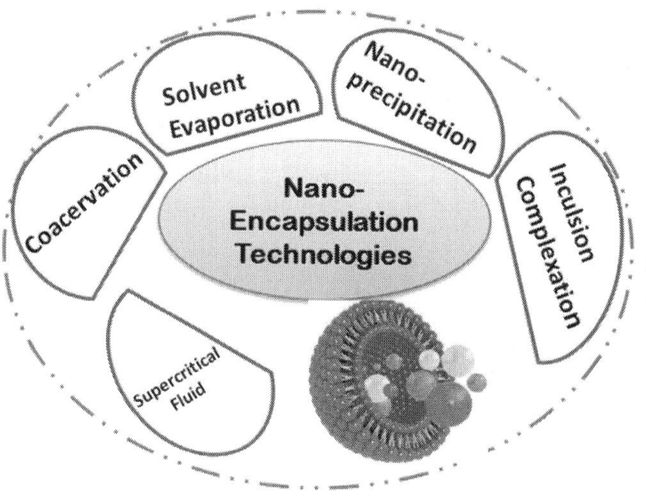

Figure 9.15. Nanoencapsulation.

Source: https://link.springer.com/article/10.1007/s10311-019-00886-w.

9.5. NANOENCAPSULATION

Nanoencapsulation (Figure 9.15) is a relatively recent method that involves encapsulating the active ingredient in another material at nanoscale sizes. To be licensed by the food industry, nanoencapsulation materials must fulfill a variety of criteria. All coating materials used for this purpose should be GRAS, protect and sustain the EO's activity, be nonreactive with the EO or food components, and be biodegradable (Ezhilarasi et al., 2012). Polysaccharides of plant origin, such as starch and cellulose, and their derivatives, different types of gums, galactomannans, pectins, soluble soybean polysaccharides, and polysaccharides of microbial and animal origin, such as dextran, chitosan, xanthan, and gellan, make up most of the coating materials. Furthermore, several lipids and proteins can be employed for this purpose. In general, there are two ways to nanoencapsulation: top-down and bottom-up. Top-down approaches are high-energy and refer to the construction of nano-sized materials via self-assembly and self-organization, whereas bottom-up approaches are low-energy and refer to the construction of nano-sized materials via self-assembly and self-organization. Although only a few nanoencapsulation technologies can encapsulate hydrophilic molecules, all methods can encapsulate lipophilic molecules. Producing lipid-based nanoparticles or polymer-based nanoparticles are the two types of procedures used to nanoencapsulated EOs. Because EOs are lipophilic chemicals, lipid-based methods are used to create nanoscale EOs for food preservation.

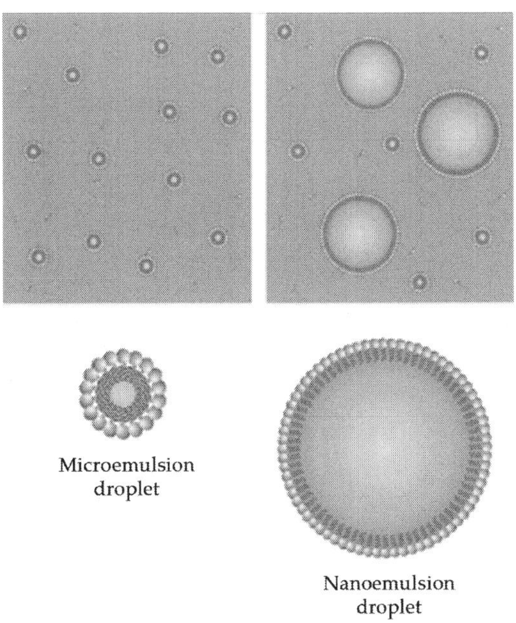

Figure 9.16. Nanoemulsions.

Source: https://link.springer.com/article/10.1007/s10311-021-01216-9.

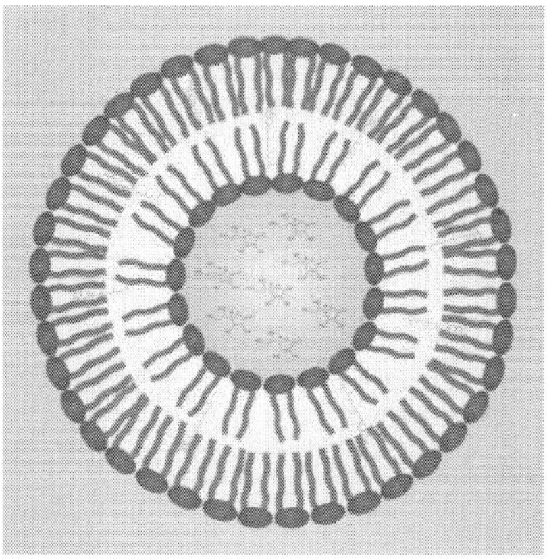

Figure 9.17. Nanoliposomes.

Source: *https://www.sciencedirect.com/science/article/pii/B9780128094365000033.*

9.6. NANOEMULSIONS

Nanoemulsions (Figure 9.16), nanoliposomes (Figure 9.17), solid lipid nanoparticles, and nanostructured lipid carriers are all lipid-based nanoparticles. Nanoemulsions are liquid-in-liquid dispersions having diameters of less than 100 nanometers. Nanoemulsion preparation can be done in two ways: top-down procedures such as high-pressure homogenization, microfluidization, or ultrasonication, or bottom-up procedures such as phase inversion and spontaneous emulsification (Helder et al., 2011). Another type of nanoscale delivery device for EOs is nanoliposomes, which are spherical lipid vesicles with an aqueous core and amphiphilic lipid bilayer. Nanoliposomes have the benefit of fusing easily with bacterial membranes and entering the bacterial cell, but their main downside is their stability.

Coating nanoliposomes, on the other hand, improves their stability and hence extends their half-life. Techniques for preparing SLNs and NLCs include hot homogenization and cold homogenization. Solid lipid nanoparticles (SLNs) are made up of fatty acids, triglycerides, steroids, partial glycerides, and waxes. Nanocapsules and nanospheres are two types of polymeric nanoparticles. Nanocapsules are made up of a polymeric shell and active agents, such as EOs, that can be put inside the core or adsorbed on the surface to create a reservoir-type nanomaterial. Nanospheres are matrix-type systems where the EO is uniformly distributed throughout the structure.

The importance of wild game foods cannot be overstated. They are distinguished by having the same nutritional content as other forms of meat produced from common sources such as domesticated species, hence providing a multisource nutritional value that is beneficial to human nutrition. However, additional precautions must be taken when preparing such foods, as they are obtained in a different way by hunting in remote, unpopulated locations with limited facilities. They have a low-fat level while having a high protein level. Furthermore, given the natural distribution of contaminants in the hunting region or in the animals themselves, including hazardous germs like Salmonella and Escherichia coli, the quality of game meats is determined by the rapid cleaning process and means of preservation, whether it by refrigeration and freezing or otherwise, after hunting them. Game meat quality is influenced by the amount of effort game animals put out to flee while being pursued before being hunted. As a result, unlike animals slaughtered under normal conditions, game animals do not bleed after being slaughtered, resulting in bigger volumes of blood staying inside game animals and hence a lesser amount of glycogen. In comparison to typical beef, glycogen serves as a source of energy in muscles, lowering their resistance to damage, deterioration, and quality preservation. With this in mind, game meats will quickly deteriorate if proper safeguards are not taken. If abattoirs can be discovered in the hunting region and meat can be carried to them without difficulty, it is critical to ensure the safety of meats and their correct processing.

The safety of wild game meat intended for human consumption is mostly dependent on hunters' and game meat dealers' awareness of good health practices and adherence to food safety fundamentals, as well as the use of sound meat conservation methods comparable to those used for other forms of meat. Prior to beginning the processing of carcasses, and when transferring from one activity to another during and after the processing phases, as well as anything that could result in contamination, adequately clean and wash hands with water and soap. Start work with clean equipment and containers that are washed and cleaned before and after each use. After each usage, wash, and clean instruments such as knives and crushing tools and devices used to remove horns such as saws and cleavers, using hot water at 82°C or disinfecting solutions. It is recommended to use chloride-containing water or to place them in a suitable container with disinfecting solution. They should be allowed to air dry. It is also a good idea to have a sufficient supply of disinfecting solution on hand for usage during the hunting. When preparing meat for preservation, it should be kept at a safe distance from other foods. To reduce cross-contamination, it is best to use distinct colored cutting boards for meat and other items.

The appropriate equipment, containers, and supplies for the processing and cutting of meat should be accessible at the location for proper carcass and meat processing. Furthermore, it is critical to employ materials that prevent the rotting of meat acquired from a variety of species, whether large or medium-sized, such as deer. People with the qualifications and experience needed to prepare and process carcasses using sound procedures and preserve and package such carcasses in a way that assures adequate cooling or freezing should be accessible within the group of hunters as part of the hunting team. This would aid in the preservation of meat after hunting for longer periods of time. Hunters should process and remove the meat as soon as possible and transport it to frozen storage, since this is the most important step in ensuring the safety and quality of the meat by reducing the risk of contamination and rotting and slowing the spread of bacteria. Preserving the carcasses of birds and small animals in their entirety is important. Unless storage bins with enough capacity are available, large game animals like bullocks or larger require the separation and chopping of smaller pieces into four sections or less. When hunting huge animals, it is best to approach them from behind after they've fallen down to ensure they're unable to move, and then control and slaughter them if they're still alive to get the most blood. Also, hunting experience, such as firing multiple shots into the neck, would allow for the most possible bleeding and hence the quickest possible death of the animal prior to approaching and slaughtering it, and hence the quickest possible death of the animal after losing the most blood.

Dragging animals out of muddy or contaminated places to a dry, clean location before skinning and removing the intestines while keeping the head and neck is crucial. To preserve the quality of wild game meat, the temperature of the animals should be changed as soon as possible. Rapidly opening the chest and bell cavities can hasten the loss of warmth. To maintain the cavity open, a clean instrument in the form of a stick might be used. Using clean water, wash the regions of the body that have been polluted by sand, clay, or grass as soon as possible. Carcasses should be fastened by one hind leg and hanged in an upright position on bars or carriages properly secured as carriages for hanging, using plastic ropes. Contamination of meat should be prevented throughout the skinning, gut emptying, or cutting process. It is vital to always use clean gloves when working and to replace or clean them after each use.

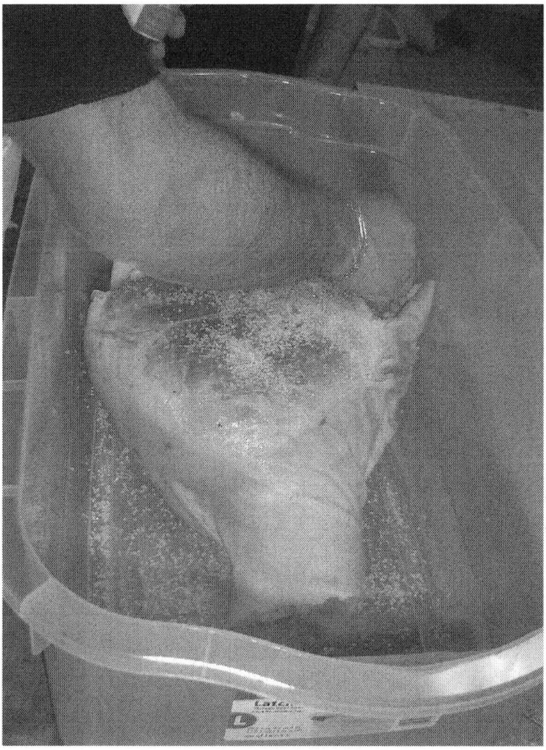

Figure 9.18. Curing.

Source: https://en.wikipedia.org/wiki/Curing_(food_preservation)#/media/File:ProsciuttoSeaSalt.JPG.

Curing (Figure 9.18) can be done in two different ways just after beef has been washed and stored in the fridge: using salt and sugar or infusing the meat with a saline solution; however, it is recommended to buy commercially produced solutions for this purpose. A multihole hollow needle should be used to infuse the saline solutions into the beef slices. It is critical to avoid creating air pockets because of incorrect injection (Joseph et al., 2009). The goal of the injection is to spread the conserving saline solution in the body tissues of the meat, as injection aids in spreading the saline solution in two dimensions: deep layers at the start of the injection, progressively till the external layers while removing the needle. While removing the needle at the end of the injection procedure, the saline solution is spread in two directions: from the inside toward the top layers, and from the surface layers toward the deep layers. As a result, the meat cut contains an equivalent amount of saline solution. The injection tool must be cleansed with soap and disinfectants between uses and should not be exposed to dirt or impurities. During the injection, the needle should be loaded with the solution and slowly injected into the meat by inserting the needle. When filling the needle with the solution, be careful not to leave any air bubbles. The cured, pickled, or dried meat is smoked at a minimum of 48°C. The smoking method seeks to give the meat a specific flavor by smoking it with certain types of dry wood to avoid any unfavorable changes in taste. The smoking procedure should be continued until the internal temperature of the meat reaches 70°C. The meat can be frozen to keep it fresh for a longer time. Wrap the smoked meat in a clean, light-colored towel.

Inoculation of contaminating microorganisms in the meat batter is made easier by the process activities used to make minced meats. As a result, foodborne pathogens in minced meats should be monitored for presence and multiplication. A single addition of nisin to L. monocytogenes injected into minced buffalo meat increased the lag period, according to one investigation. In minced meats, combining bacteriocins with plant essential oils at levels that do not impart an unpleasant flavor is being studied as a means to boost L. monocytogenes inactivation and Salmonella enteritidis inhibition. Nisin's antilisterial effect in minced beef was considerably enhanced when it was combined with thyme essential oil. During storage at 4 °C for at least 12 days, the combination of 0.6% essential oil and 1,000 IU/g nisin reduced the population of L. monocytogenes below the statutory European Union limit. The thyme oil did not impart an unpleasant flavor at that concentration. Inhibition of S. enteritidis in sheep minced meat by a combination of nisin and oregano essential oil has also been reported as promising, although nisin at 500 or 1,000 IU/g had no effect against S. enteritidis when used alone. The combination of oregano essential oil at 0.6% and nisin at 500 IU/g had better antibacterial activity against S. enteritidis than the single oregano essential oil at 0.6%, but was less effective than the combination of oregano essential oil at 1,000 IU/g.

Cooked meat products are one of the most popular ready-to-eat foods on the market. They can be produced using complete chunks of primary meat, but most are prepared by grinding and mixing secondary meats or blood with other ingredients, then stuffing and cooking. Natural microbiota is inactivated during the cooking process, allowing postprocess pollutants to flourish. Most cooked meat products have pH levels that are conducive to the growth of pathogenic and spoilage bacteria, which can flourish at low temperatures over the product's shelf life. Further processing, such as slicing, peeling, and packaging, may be applied to some of these meats, increasing the potential of cross-contamination. For these reasons, the use of bacteriocins as barriers against spoilage bacteria and diseases has piqued interest. The major options investigated were adding bacteriocin preparations to meat slurries before heating, applying bacteriocins to the surface of the meat before packaging, and applying bacteriocin-dosed films or coatings. Because of their thermotolerance, the potential of putting bacteriocins in the meat before cooking is intriguing (FAO, 1990).

Chapter

10 Advances in Biopreservation

CONTENTS

10.1. INTRODUCTION

We live in a society where two interconnected global concerns exist: the energy and food crises. Neither can be solved satisfactorily unless the other is addressed. To take the argument to its logical conclusion, efficient global food production can only be achieved with the help of artificially created energy, and there will be no need for such energy until mankind has the food it needs to survive. In both concerns, the IAEA has an interest and a responsibility.

Its initiatives in these domains are focused on using the atom to produce energy and improving the global food crisis. The world food problem may be addressed in two ways, first by expanding food production, that is, growing more food using agricultural expertise and material input, and by preserving more of the food currently produced using food science and technology.

While both paths should be pursued, it seems obvious that preservation should take precedence, not only because it would reduce wasting of existing food that is critically required by millions of people, but also because it would prevent the waste of energy invested in food production. Food preservation is thus a critical technology, and research into both old and innovative ways of food preservation should be pushed aggressively.

Irradiation of food, usually with gamma rays from a cobalt-60 source, has several advantages over traditional methods. It can not only delay the onset of undesirable physiological changes, microbial spoilage, and insect damage, but it can also kill disease-causing organisms that will infect the food if it is not irradiated.

It's astonishing to learn that, after 25 years of extensive research, the practical implementation of food preservation by irradiation has only recently begun in one IAEA Member State. The main roadblocks are psychological in nature, but there are a number of other issues to be resolved as well. Despite this, it is evident that many questions have been answered, and great progress has been achieved.

A few of these developments will be listed shortly below. However, it should be highlighted right once that, like traditional approaches, irradiation has limitations and cannot be considered a panacea for all food concerns in all circumstances.

10.2. BIOPROTECTION

The use of Bioprotection measures in food preservation may be limited by national laws that range significantly in its basics and practical implications. Some are clearly relevant to the addition of bioprotectants to foods, but others may be related to seemingly unrelated issues like labeling, packaging, export, the use of biological agents, or genetically modified organisms. The process of obtaining a license to use antimicrobial compounds as food preservatives is possibly the most difficult. Nisin is currently the only bacteriocin approved for use as a food preservative). Nisin is allowed in more than 80 countries. National laws regulating the presence and amount of nisin in food products, on the other hand, varies significantly. Nisin, known as Nisaplin™ (Figure 10.1) and Chrisin™ (Figure 10.2) is a lyophilized substance produced by microbial fermentation. The products of industrial fermentation are governed by normal food rules. Fermentation concentrates or lyophilized powders can be used as components in foods.

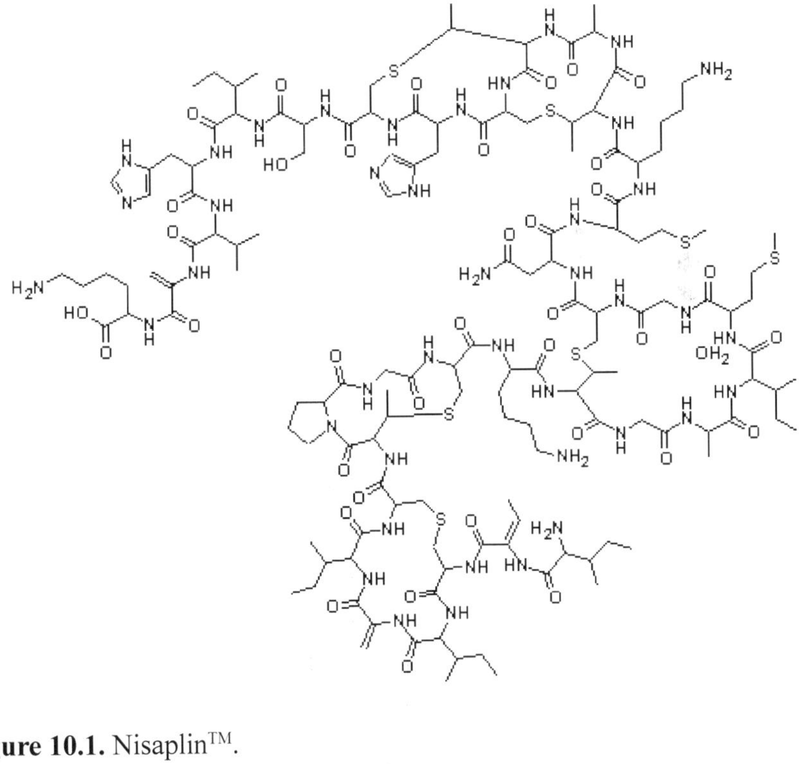

Figure 10.1. Nisaplin™.

Source: http://images-a.chemnet.com/suppliers/chembase/cas/cas1414-45-5.gif.

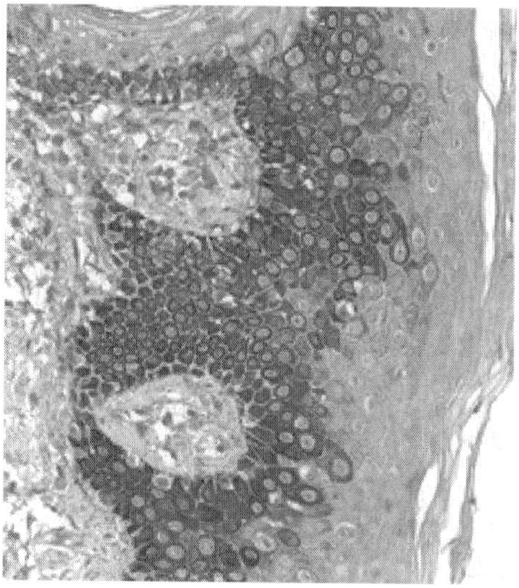

Figure 10.2. Chrisin™.

Source: https://www.sigmaaldrich.com/catalog/product/sigma/r3251?lang=en&region=IN.

Fresh produce products can be contaminated with human pathogenic bacteria from a variety of sources and have been linked to several outbreaks. Several bacteriocin preparations like nisin, pediocin, or enterocin AS-48 have been tested on the surfaces of fresh-cut vegetables and sprouted seeds for inactivation of foodborne pathogenic or toxicogenic bacteria. Bacteriocin treatments have also been proposed for decontaminating whole fruit surfaces, preventing pathogenic bacteria transmission from fruit surfaces to processed fruits, and reducing bacterial survival on sliced fruit surfaces during storage.

The addition of bacteriocins like nisin to vegetable fermentations has been proposed as a way to direct the microbiota towards selection of desirable bacteriocin-tolerant or bacteriocin-resistant strains with desirable effects while inhibiting strains that cause defects such as overripening. The addition of a nisin preparation to cabbage infected with nisin-resistant L. mesenteroides improved fermentation control and delayed LAB development. Nisin was added to kimchi to control lactobacilli that cause the product to over ripen. Lactobacillus spp. were shown to be more resistant to Nisin than Leuconostoc spp.

B. subtilis is the most common cause of ropiness in bread, though B. licheniformis, Bacillus megaterium, and Bacillus cereus may also be implicated. Rope formation can develop in acidified wheat loaves, as well as loaves with high sugar, fat, or fruit contents. Nisaplin or nisin-producing lactic acid bacteria have been found to be ineffective in inhibiting B. subtilis and B. licheniformis strains in bread making. B. subtilis rope formation was inhibited by fermented broth from a Lactobacillus plantarum strain generating bacteriocin-like inhibitors in yeast-leavened bread. Other bakery goods, such as refrigerated pizza, may contain a range of ingredients, making microbial deterioration management more difficult. Due to the suppression of spoiling lactic acid bacteria, Nisaplin under modified environment packing significantly enhanced the shelf life of ham pizza.

The generation of bacteriocin by bacterial species is of particular importance for the prevention of harmful germs in seafood. For the competitive exclusion of L. monocytogenes in processed food products, antagonistic bacterial strains could be used. Many LAB strains have the ability to grow at low temperatures. They can withstand a modified environment, low pH, high salt concentrations, and additions like lactic acid, ethanol, or acetic acid. The antagonistic strains chosen are able to grow on the food product during cold storage and create antimicrobials to inactivate or at least impede the growth of L. monocytogenes and should have no negative effects on the food product. They should not have any negative impacts on health or carry antibiotic resistance or virulence features. Although inoculated strains may have probiotic qualities, this method of probiotic administration through marine products has yet to be explored.

In order to extend the shelf life of liquid whole eggs and inactivate foodborne germs, they must be processed for commercial usage. Ultra-Pasteurization was devised as an alternative to standard pasteurization, resulting in a liquid whole egg with a shelf life of at least 10 weeks at 4°C when combined with aseptic processing and packaging. In order to produce ultra-pasteurized liquid whole eggs with an extended pathogen-free shelf life, effective aseptic filling and packaging technologies are still required. Listeria monocytogenes is not completely inactivated by ordinary minimal egg pasteurization methods, unlike Salmonella. Listeria species can, in fact, be isolated from commercially broken raw liquid whole eggs. As a result, it was suggested that bacteriocins be used to manage Listeria in this food system. The addition of nisin to pasteurized liquid whole eggs reduced viable counts of L. monocytogenes, improved product refrigerated shelf-life, and protected the liquid egg from L. monocytogenes and Bacillus cereus development during storage.

By using the casting method, a whey protein edible film containing an ethanolic extract of Fucus vesiculosus was successfully created. The F. vesiculosus extract appears to reinforce some of the physical features of whey protein films, such as thickness, tensile strength, and elastic modulus. The extract, on the other hand, rendered the film more permeable to water vapor. In comparison to the control whey protein film, the novel film was able to prevent lipid oxidation in packaged poultry meat for at least 25 days of storage. The inability to perform further lipid oxidation assays, such as the identification and quantification of the aldehyde hexanal, which is responsible for the rancid smell and taste, was one of the study's shortcomings. Furthermore, the acquired results urge future assays, such as microbiological assays and sensory analysis, to assess the influence of this novel film in extending the shelf-life of poultry meat.

This study emphasizes the effects of fennel essential oil (FEO) and peppermint essential oil (PEO) on chitosan-based films. The addition of FEO and PEO raised the density and thickness marginally while lowering the moisture content, water swelling, and solubility characteristics dramatically. The color values of composite films containing FEO and PEO altered noticeably with a tendency toward yellowness, which was useful in resisting food breakdown caused by ultraviolet light. The addition of FEO and PEO changed the structure of the chitosan films, while the interaction between the chitosan and polyphenols in FEO and PEO generated new hydrogen bonds and increased the thermal stability, according to the differential scanning calorimetry (DSC) and Fourier-transform-infrared (FTIR) data. The surfaces of the composite films comprising FEO and PEO were smooth, but the cross-section was rougher than in pure chitosan film, according to environmental scanning electron microscopy (ESEM).

The bacteriocin EFL4 produced by the carcinogenic E. fecalis L04 strain can be employed as a natural biopreservative for RTE fresh salmon filets during cold storage. The MIC for S. putrefaciens was 0.32 g/mL in this investigation, indicating that the bacteriocin EFL4 could suppress some spoilage bacteria development. This component was refined and used to help keep RTE fresh salmon filets fresher for longer. The microbiological, physicochemical, and organoleptic features of RTE fresh salmon filets treated with 0.64 g/mL showed that they kept their quality better during cold storage, which is likely because bacteriocin EFL4 efficiently inhibited the growth of fish-spoilage bacteria and foodborne pathogens. In general, bacteriocin EFL4 can be considered a viable and effective fish biopreservative; however, more research should be done to optimize bacteriocin manufacturing and large-scale procedures in order to increase the fish industry's economic profitability when employed as part of hurdle-based preservation technology.

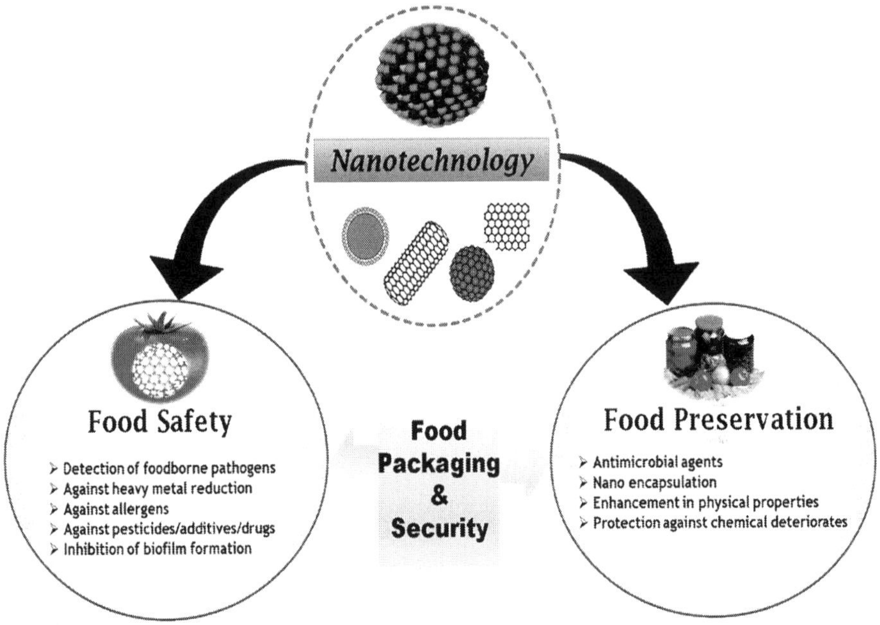

Figure 10.3. Nanotechnology.

Source: https://ars.els-cdn.com/content/image/1-s2.0-S1021949818301169-fx1_lrg.jpg.

Nanotechnology (Figure 10.3) is becoming more prevalent in the food business, and several researches have been conducted in recent years. Because of the complexity of the technology and the food matrix, the application of nanotechnology to deliver natural antibacterial chemicals in meals is limited. Egg-white lysozyme with silver nanoparticles is effective against E. coli, Bacillus anthracis, Staphylococcus aureus, and Candida albicans while maintaining the enzyme's hydrolase function. These nanoparticles were effective against silver-resistant Proteus mirabilis strains as well as an antibiotic-resistant and silver-resistant E. coli strain. These nanoparticles were shown to be nontoxic at the quantities utilized to prevent microbial growth in human epidermal keratinocytes. Nisin nanoparticles, as well as bacteriocin nanovesicles have been tested against L. monocytogenes and S. aureus with promising results. The bacteriocin's antimicrosporidia action was amplified, resulting in a sustained decrease in fecal spore shedding and intestinal spore load.

Because consumers are increasingly demanding food that is free of synthetic preservatives, new alternatives must be identified and studied. These new methods should help limit foodborne infections while also prolonging the shelf life of foods. The search for natural antimicrobials must be cost-effective from an economic standpoint, and one alternative approach would be to combine several natural antimicrobials with food preservation techniques. Natural antimicrobial chemicals may bind with some dietary components, limiting their effectiveness, due to the complexity of food matrices. Although this technique has prompted concerns about consumer safety, nanoparticles and/or nanovesicles have immense potential in food safety as an effective antimicrobial delivery mechanism. As a result, more research is needed to determine the most effective antimicrobial delivery technology as well as the optimal concentrations of natural antimicrobial compounds.

Food spoilage is a serious problem in the food industry, resulting in wasted food, significant financial losses for both manufacturers and consumers, and a bad influence on brand names. Fungal contamination is one of the most common causes, and it can occur at any point in the food chain. Fungal growth causes food sensory flaws ranging from visual degeneration to significant odor, flavor, or texture abnormalities, but some molds also produce mycotoxin, which can have serious health consequences. Various treatments, such as fungicides and chemical preservatives, are employed to prevent microbial deterioration and hence improve product shelf life. Simultaneously, government officials encourage the food business to reduce the usage of these chemical substances and create natural food preservation alternatives. This is coupled by a huge societal demand for 'clean label' food goods, as customers seek for products that are more natural, less processed, and safer. In this context, antifungal microbial agents, such as bioprotective cultures, fermenters, culture-free supernatant, or isolated compounds, are gaining popularity as a viable alternative to chemical preservation.

10.3. HURDLE TECHNOLOGY

Furthermore, hurdle technologies are used in accordance with the process and raw materials. They can include thermal treatments like pasteurization, sterilization, and heating at ultra-high temperatures; water removal techniques like drying, freeze-drying, and smoking; water activity reduction techniques like adding high sugar and salt concentrations; pH modification techniques like immersion in vinegar or other acids; and nonthermal preservation techniques like modified atmosphere packaging. To avoid fungal spoilage, chemical preservatives such as benzoate, propionate, sorbate, nitrate, nitrite, and sulfites are utilized. Natamycin is the only antifungal chemical produced from bacteria that is utilized as a food preservative. Streptomyces nataliensis produces it, and it belongs to the polyene macrolide antimycotics family. Most nations have approved the use of natamycin as a surface treatment for cheese, and its inclusion in other foods is subject to regulations. Because of the possible negative environmental and health impacts of certain fungicides and preservatives, more natural alternatives have emerged. In the case of crops, unwanted residual pesticide concentrations on harvested products, regulations, acquired fungal resistance, and environmental effects have led to the development of biocontrol agents, which is currently a fast-growing sector in need of natural remedies to reduce crop losses. Several bacteria, fungus, and yeast-based products are already available for purchase globally.

Due to their natural nutrient-rich nature, food products are sensitive to fungus deterioration. However, susceptibility is dependent on several factors, including the nature of the food matrix and biological, physical, and chemical parameters and management during fruit and vegetable harvesting and raw material storage. The intrinsic properties of the food matrix, as well as associated external influences, will influence the growth of specific fungal genera or species. On a particular matrix, this usually results in the selection or domination of one or more fungal species. In terms of quality, fungal contaminations can cause visual and/or other flaws. The most prominent is apparent fungal development due to the presence of thallus or yeast colonies, which finally results in rotting. Fungal development can also be linked to the occurrence of black, white, green, pink, or yellow patches. At the industrial or consumer level, this usually results in the complete removal of the product. The effects may be less obvious to detect depending on the nature of the commodity and the type of contamination, but fungal metabolism can cause a variety of organoleptic defects such as gas production, off-flavors, and texture alterations.

In terms of health risks, fungal growth and presence can lead to the release of mycotoxins, which can be produced by species belonging to the Aspergillus, Penicillium, Fusarium, and Alternaria genera. This is significant since fungi and mycotoxins are expected to infect around 25% of agricultural raw materials worldwide. At low quantities, several mycotoxins can cause toxic, carcinogenic, and mutagenic effects in humans, with aflatoxins, ochratoxin A, patulin, fumonisins, zearalenone, T-2 and HT-2 toxins, and deoxynivalenol being the most prominent. The fundamental issue is the existence of mycotoxins in processed foods, which can no longer be identified, but these metabolites are usually resistant to technological remedies.

10.4. MYCOTOXINS

As a result, mycotoxin concentrations have been regulated in several countries. Aflatoxin A and aflatoxins are major mycotoxins in food spoilage, and ochratoxin A is produced by numerous Aspergillus species, whereas OTA is generated by several Penicillium species. The principal aflatoxin generator, Aspergillus flavus, is a common spoiler of a variety of foods. Fumonisins are produced by the fungus Fusarium and the fungus Aspergillus (Fun et al., 1991). Fumonisins and ochratoxins were produced by Aspergillus awamori strains infecting onions and Aspergillus niger strains discovered in coffee beans and grapes. Furthermore, items manufactured from raw materials containing these mycotoxins provide an additional risk. Patulin is generated by many Alternaria, Paelomyces, Aspergillus, and Penicillium species, and is noncarcinogenic yet immunotoxic and neurotoxic in animals. Penicillium expansum (Figure 10.4), a common apple spoiler, can release patulin, which can subsequently be transferred to juices and other derived goods.

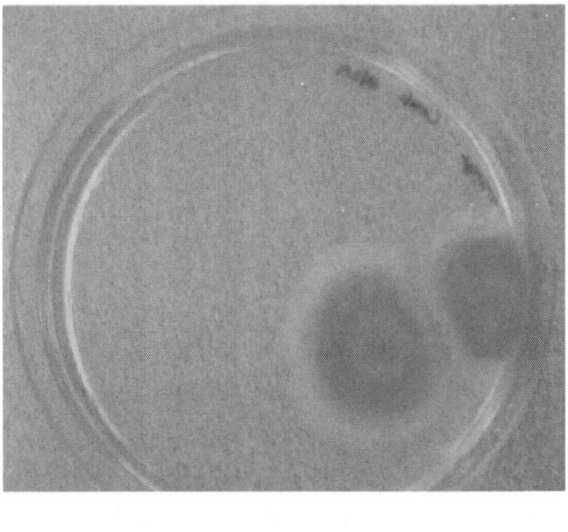

Figure 10.4. Penicillium expansum.

Source: https://en.wikipedia.org/wiki/Penicillium_expansum.

10.5. FOOD PACKAGING

Food packaging is one of the most important measures in ensuring food safety. No packaging materials are impervious to natural chemicals, air gases, or water vapors. In the case of packaging fresh fruits and vegetables that undergo cellular respiration, however, completely preventing gas migration and permeability is not ideal. Carbonated beverage packaging, on the other hand, should block the flow of oxygen and carbon dioxide to avoid oxidation and decarbonation.

Depending on the food matrices and packing materials utilized, the flow of CO_2, oxygen, and water vapors varies. As a result, numerous nanocomposite materials, including polymers, can be used to address and overcome these challenges in food packaging.

A nanoparticle with a diameter of less than 100 nm is thousands of times thinner than a book page with a thickness of 100,000 nm or a hundred times thinner than a human hair with a thickness of 10,000 nm. Different shapes of nanoparticles can be of substantial utility in numerous domains of the food and pharmaceutical industries, notably for research programs related to food science-based research, with the proper regulatory processing approval. The usage of nano-biocomposites in food packaging has recently improved food packaging's ability to operate as a gas barrier.

A nano-sized titanium dioxide or tin dioxide particle-based advantageous photo-indicator intelligent ink employing a redox-activate methylene blue dye for in-packaging oxygen monitoring was prepared, and this detector progressively changed color in response to tiny changes in oxygen quantity. Because of their easy accessibility, low cost, easy processability, and high performance, the packaging industry has focused on various types of nanostructures and specific nanomaterials, including nano-clay particles, in recent years.

In addition, graphene nanosheets and carbon nanotubes are being created as carbon-based nanomaterials with promising results. Due to their capacity to block out oxygen and odors, polymeric substrate-based flexible or static films and bottles in combination with vacuum-coated thin inorganic layers are in high demand in material packaging. While aluminum foil and aluminum metallization were once the preferred materials and methods, new materials are now being employed more frequently.

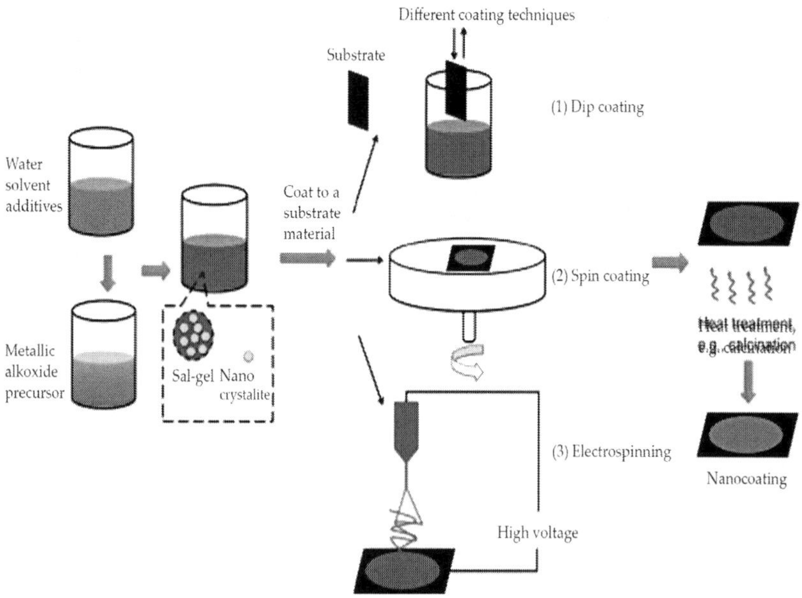

Figure 10.5. Nanocoatings.

Source: https://www.sciencedirect.com/topics/engineering/nanocoatings.

10.6. NANOCOATINGS

The term "nanocoatings" (Figure 10.5) refers to thick and uniform layers of certain particles that are at nanoscale. Several researchers have produced an intelligent nanocoating film that can detect contamination during storage in the last few years. Gas content and noninvasive detection technologies have also demonstrated a strong ability to monitor the gas content, excess moisture, and oxygen content of a package-headspace continuously and simply, allowing for effective evaluation of food quality and safety even after the manufacturing process.

Due to oxygen's propensity to establish a healthy environment for microbial development, the scarcity of oxygen inside the packaging might shorten the shelf life of food (Taehyung et al., 2019). Nanocomposite of silica in-situ to construct biodegradable and cost-effective polyvinyl alcohol/ chitosan polymer-based biodegradable and cost-effective films resulted in significant reductions in oxygen and moisture permeability of 25.6% and 10.2%, respectively, implying that the films prolonged the preservation duration of cherry by three-fold as compared to the usual packaging.

Magnesium oxide (MgO) nanostructures and polylactic acid biopolymer can be used to produce food packaging material that effectively defends against bacterial biofilms.

The innovative usage of antimicrobial nanoparticles has helped to prevent our food products from foodborne outbreaks caused by the ingestion of damaged packaged food to some extent. When compared to traditional food packaging, an active food packaging not only acts as a passive barrier, but also aids in the release of antioxidants and antibacterial chemicals through direct contact with the food, as well as the removal of some unfavorable components like oxygen or water vapor. Food stability is frequently improved by these interactions. Specific molecules in active packaging have the potential to absorb or release components into or from the packaged food's surrounding environment. Currently, active polymer nanoparticles in food packaging are mostly used for antimicrobial packaging. Overall, numerous bioactive chemicals can be included in packaging material through capsulation, capping the material with nanoparticles, or using other nano techniques to increase the functionality and effectiveness of food packages. Food providers should choose effective antimicrobial substances-based unique approaches to ensure the quality of packaged products, such as their visual appearance and sensory levels, in order to properly use antimicrobial active packaging for food goods. Nano-encapsulation permits nanomaterials to come into close touch with the human body while eating. Silicon dioxide, for example, is widely used as an odor carrier in a variety of culinary products. Several lipid-based nano-encapsulation methods have been developed to improve the bioavailability and solubility of antioxidant components, as well as their targeted, site-specific delivery with effective absorption. Nano-sized edible coatings have been demonstrated to be an excellent alternative for food preservation and shelf-life extension, as well as preventing microbiological deterioration. Coatings produced with gelatin integrating cellulose nanocrystals, chitosan or nano-silica, chitosan with nano-silica, and alginate or lysozyme nano-laminate have been demonstrated to be particularly successful in preserving the freshness of fresh meals during long-term storage. Furthermore, a nano-packaging-based technology incorporating blended polyethylene with nano-powders such as Ag, kaolin, and rutile TiO_2 has been used as a unique and simple way for preserving fruits like strawberries. Despite extensive study into nano-encapsulation utilizing a variety of materials, the exact delivery of foods and their safety have yet to be thoroughly investigated. As a result, future studies should focus on the toxicity of nano-encapsulated meals over time.

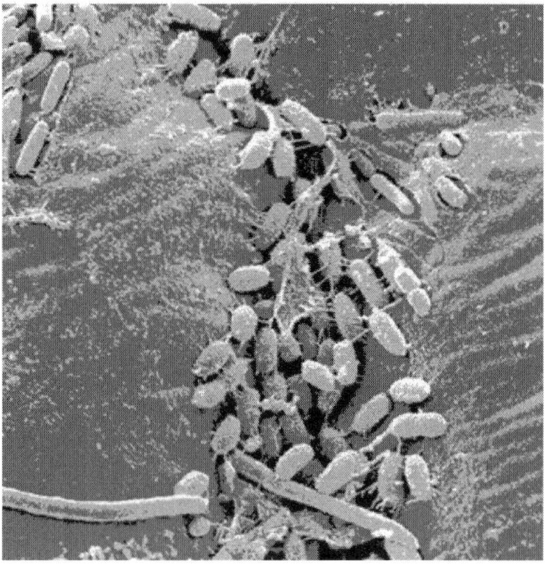

Figure 10.6. Biofilms.

Source: https://www.food-safety.com/articles/4137-beating-back-biofilms-in-food-processing.

Biofilms (Figure 10.6) are clumps of bacteria that stick to a variety of surfaces and generate a polymeric extracellular matrix that is tough to penetrate. Biofilm production begins with the attachment of free-floating microbes to the surface via van der Waals forces, which cause challenges in the food processing industry such as biocorrosion, biofouling, and accumulation. Glycerol monolaurate or GML, which has been approved by the US Food and Drug Administration, is an antibacterial drug that works against a variety of Gram-positive bacteria, including Bacillus anthracis. GML has been found to suppress the production of biofilms in three different S. aureus strains as well as MRSA (Amy Wong, 1998). In addition, antimicrobials are used throughout the nano-fibers of filter membranes to avoid biofilm formation. Nickel oxide nanoparticles or NiO NPs have also been suggested as antibacterial and antitumor agents. Biofilm inhibitory effects of chlorhexidine-conjugated gold nanoparticles against Klebsiella pneumonia. Bacillus subtilis has shown significant potential in the production of value-added products in a variety of industrial processes. The ability of B. subtilis to create biofilm is strongly tied to its fermentation process, resulting in operational challenges and important processes.

Bacteriophages are highly common and diverse in nature. In the biosphere, there are over 1031 phage particles. Phages are obligate intracellular parasites with no intrinsic metabolism that multiply in metabolizing host bacteria. They are nonpathogenic to humans. Following adsorption, phages inject their genomes into the bacterial cytoplasm and multiply via one of two primary life cycles: lytic or lysogenic. They cause bacterial death during the lytic cycle by disrupting the host cell and releasing the phage progeny. Phages bind to specific sites on the cell wall, and their host range is determined by this specificity. Bacteriophages have been used for a variety of applications in clinical and veterinary medicine, microbial source tracking for water evaluation, and in the agro-food business. Apart from its use in food safety for pathogen reduction in living animals during primary production and disinfection of surfaces and equipment that come into contact with processed food, bacteriophages could also be employed as a bio-preservation method.

10.7. PHAGES

Phages used to preserve ready-to-eat food can have a wide host range. Polyvalent phages or broad host range phages can infect a wide range of target strains, species, or targets. Bacteriophage host range varies through time and can demonstrate unexpected adaptability across species, genera, families, and even classes boundaries. Severe microbes can be transmitted to humans through fresh-cut vegetables and ready-to-eat items, resulting in foodborne diseases and even death. Salmonella enterica, Listeria monocytogenes, and Shigella are only a few of the dangerous bacteria that can be found in ready-to-eat fruits and vegetables. Furthermore, due to the increasing number of immune-compromised individuals and the particular processing needs for ready-to-eat food and fresh-cut fruits and vegetables prior to commercialization, food companies are experiencing issues connected to consumer sensitivity to foodborne pathogen infections. Furthermore, ready-to-eat food is highly perishable and cannot be preserved using typical processing methods such as heat, pasteurization, or dehydration, as these processes may alter organoleptic qualities and lower nutritional value.

Microbes that cause food rotting represent a serious threat to people. Spoilage causes unwanted changes in the sensory properties of food, rendering them unfit for human ingestion. Microbial deterioration is the most common type of food decomposition. Microbial deterioration results in the loss of about 25% of all food produced each year. Even when existing preservation procedures, excellent production procedures, quality control, and hygiene standards are followed, food loss happens. Food firms must continuously exceed customer expectations for pathogen-free, high-quality products that do not use artificial preservatives. As a result, efforts are being made to prevent microbial contamination by using natural antimicrobials. Bacteriophages are a type of natural antibacterial with well-known biotherapeutic potential all around the world. The approval of the first phage product, ListShield TM, to control Listeria monocytogenes in meat and poultry products, was a major milestone in the western world's phage history, which occurred in 2006. Following that, phage products were recommended for use as biotherapeutics in meals. Biological antimicrobial systems, often known as biopreservation, are among the new and creative technologies being explored throughout the world. Biopreservation is a method for preserving food and extending its storage life that involves the use of natural or controlled microflora as well as antimicrobials. Food biopreservation involves the use of lactic acid bacteria, which prevent food deterioration germs. Beneficial bacteria or bacteria themselves create fermentation products that are used in biopreservation to inactivate pathogens in food and control deterioration. Beneficial bacteria inhibit the growth of spoilage bacteria in a variety of methods, including lowering pH through organic acid synthesis and producing different undissociated acid molecules having antimicrobial action. Bacteriophages and yeast are the other biopreservative agents used.

Bacteriophages are parasitic bacteriophages that multiply and take over the host's protein machinery. Bacteriophages can infect host cells in one of two ways: lysogenic or lytic. Genetic material is integrated into the bacterial chromosome without cell death in the lysogenic process, and newly generated virus is discharged from host cells by cell lysis in the lytic pathway. From the standpoint of food safety, using lytic phage is one of the safest antibacterial methods available. In the final stage of the lytic cycle, lytic bacteriophages release hydrolytic enzymes called lysins/endolysins, which degrade the peptidoglycan layer of Gram-positive bacteria. As biocontrol agents, bacteriophages have several advantages. Phages are preferred over other antimicrobial agents because they do not harm commensal microbiota. This allows to bacterial cell wall receptors, which have a high selectivity for target host cells.

Phages have a self-duplication and self-limiting property, meaning that a little or single amount of phage multiplies until it reaches a host threshold. Even if bacterial host cells acquire phage defensive mechanisms to ensure their survival, phages have a tendency to adapt to new host systems. Because phages are made up of nucleic acids and proteins, they have a low intrinsic toxicity. Phages are easy to isolate, propagate, and keep for a long time. They have the ability to withstand diverse physicochemical conditions as well as food processing environmental stressors. They can be found in large quantities in a variety of foods and have been isolated from raw, processed, fermented, and seafood. Phage treatment is a method of reducing pathogen colonization that is used before slaughter or throughout the growth of animals in order to reduce the risk of cross-contamination during food processing. To manage Campylobacter and Salmonella in chicken, phages are added to drinking water or food. Phages are also sprayed on chickens to control E. coli, and given orally/rectally to ruminants to control E. coli. Application of phages directly to food surfaces, such as meat surfaces, processed foods, fresh produce, or raw milk, is one method of biocontrol. Bio Sanitation entails the use of phages to destroy biofilms generated on equipment surfaces by pathogenic bacteria.

Depending on their life cycle, they might be classed as virulent or temperate. The lytic life cycle is shown by virulent bacteria, whereas temperate bacteria enter the lysogenic cycle by inserting their DNA into the bacterial chromosome or prophage. Virulent bacteriophages can lose their host bacteria and have antibacterial properties. Bacteriophages can be used to prevent or inhibit colonization, manage diseases in animals using phage therapy, decontaminate carcasses and other raw products such as fresh fruit and vegetables using phage bio-sanitation, and extend the shelf life of processed foods using phage biocontrol. Bacteriophages have been shown to be effective in reducing pathogen carriage in animal farming, as well as after slaughter or milking. Bacteriophages are efficient against Salmonella and Campylobacter in chickens and pathogenic E. coli in ruminants. Biofilm reduction can be achieved by lowering or eradicating S. aureus nasal or cutaneous colonization in food handlers. Endolysins' significance in food biopreservation is still in its infancy. However, a number of endolysins have been identified that are effective against a variety of zoonotic and foodborne pathogens. Several experts currently believe that transgenic cows that express endolysins have a lower risk of mastitis and S. aureus milk contamination. Similarly, Staphylococcal endolysin has been discovered to be effective in reducing S. aureus biofilms. Endolysins have also been linked to dairy processing, with pure endolysins being found to kill S. aureus in pasteurized milk. According to reports, recombinant LAB can release active Listeria endolysin, but their antagonistic activity in milk still must be verified. It has also been used in food safety and has a high specificity for CBDs.

Overall, bacteriocins, bacteriophages, and endolysins have a potential role in food processing, preservation, and food safety. Bacteriocins and endolysins are thought to be better suited to DNA shuffling and protein engineering in order to produce extremely powerful variations with a wider activity spectrum. Although genetically engineered bacteriophages may be useful in biopreservation, their safety concerns must be addressed before they are chosen as a biopreservative agent. Bacteriocins are a good hurdle agent that can be successfully included into packaging films and used in combination with modified atmosphere packaging (MAP). Similarly, phages and endolysins have been successfully coupled with nisin, and endolysin activity is enhanced by high hydrostatic pressure. Bacteriophages and endolysins have also been recommended for use in food contexts, including by food handlers, however delivery systems must be implemented. So, in order to achieve bio-preservation in the most holistic way possible, it is necessary to demonstrate resistance to bacteriocins and bacteriophages. Endolysins, on the other hand, has yet to be linked to such an issue. Phages, on the other hand, mutate at a much higher rate than bacteria. New phage selection, on the other hand, maybe able to quickly overcome bacterial resistance. Because of the deterioration of the lysogeny process, bacteria that are resistant to infection become superinfected. Temperate phages should therefore be avoided at all costs. Bacteriocins, bacteriophages, and endolysins are projected to be the future bio-preservatives, thanks to advances in genetic technology and molecular procedures.

Almost all foods are made from live cells derived from animals and plants, and in some cases from microbes using biotechnology. As a result, foods are mostly made up of "edible biochemicals." One of the most essential objectives of a food scientist is to make foods as safe as possible, whether they are consumed raw or cooked. Foodborne illness outbreaks can be prevented by using appropriate food processing, storage, and preservation procedures. Foodborne illness is defined as the development of sickness or illness as a result of consuming contaminated food. When billions of cans, jars, packages, and pouches of processed and fresh food products are consumed each year, the processed food industry has a record of averting such incidents. Limited outbreaks in which people succumb to the consequences of hazardous foods have occasionally shattered this outstanding record.

Food preservation is an action or method for preserving foods at a desired quality level. Several innovative preservation techniques are being developed to meet contemporary needs for cost-effective preservation as well as consumer satisfaction in terms of safety, nutrition, and sensory perception. People like eating a wide variety of foods with various tastes, flavors, nutritional, dietetic, and other properties. Unfortunately, it is believed that up to 2 billion people do not have enough to eat, and that up to 40 000 people die every day from diseases linked to poor diets, such as a lack of food, protein, or particular nutrients. In extreme situations, inadequate nutrition can cause progressive protein shortage in children, known as kwashiorkor or the more common protein deficit.

Environmental elements such as temperature, humidity, oxygen, and light induce major food deterioration processes, which can trigger a variety of reaction mechanisms that cause food to deteriorate to the point that it is either rejected or dangerous to the customer. Food degradation and spoiling are mostly caused by microbial impacts. By far the most prevalent method of food preservation is heat. The phrases used to describe the various degrees of heating preservation that eventually determine the sort of final product generated are pasteurization and sterilization. However, in order for these treatments to be effective, they must be carried out under a combination of rigorous temperature and time control to ensure both pathogenic and nonpathogenic bacteria are killed. Thermal inactivation of food enzymes and some degradation of food contents are also caused by these same processes.

Many foods naturally contain organic acids and their esters, while fermented foods include organic acids and their esters as a result of microbial activity. Many foods are preserved by the addition of low amounts of these chemicals, all of which have pH-dependent preservation activity. At low concentrations, these chemicals are predominantly active against yeast and molds, but bacteria are also impacted. Lowering the pH raises the amount of undissociated acid molecules, which boosts the antibacterial power of all organic acids. As a result, it's widely considered that these acids' antibacterial action is proportional to the concentration of their undissociated molecules. Microorganism sensitivity to weak organic acids is a crucial species-dependent characteristic. Organic acids and esters cover a wide range of compounds, although only a few are employed in food.

Food fermentation contributes significantly to the preservation of foods by producing organic acids. Many dairy products rely on lactobacilli's metabolic activities to keep spoilage microbes at bay. This is thought to be owing to the production of lactic and acetic acids, but hydrogen peroxide production could also be a component antimicrobial agent. In the case of meat, it is thought that the preservation activity is predominantly due to the lowering of pH rather than the generation of lactic acid. Flavor generation is critical in dairy fermentation. Temperature, salt concentration, and initial pH have all been found to influence the rate of acid formation. It has been proposed that the presence of glucose in meat plays a crucial role in the rate of rotting of meat.

The Na and K salts of nitrite and nitrate are commonly utilized in the fermentation of meat products and the curing of pig during the production of ham and bacon. These chemicals are significant because they stabilize the red meat color and limit the growth of harmful and spoilage microorganisms. They were originally added combined with sodium chloride. Many bacteria convert nitrate to nitrite, which aids in the prevention of microbial deterioration. As the pH is reduced, the antibacterial efficacy of nitrite increases. C. botulinum development is inhibited by nitrite, which would otherwise provide an unacceptable risk in such products. Nitrite also helps to keep cured foods from becoming rancid.

Ionizing radiation has both direct and indirect impacts on biological materials. Chemical events occur as a result of energy deposition by the radiation in the target molecule. The indirect effects are caused by reactive diffusible free radical forms such as the hydroxyl radical (OH), a hydrated electron, hydrogen atom, hydrogen peroxide, and hydrogen that result from the radiolysis of water. The hydroxyl radical is a strong reducing agent, while hydrogen peroxide is a strong oxidizing agent and a poison to biological systems. These two radicals can alter the molecular structure of organic things, including foods, in a variety of ways. Microorganisms are inactivated by a shock wave generated by an electric arc, which causes highly reactive free radicals to develop from diverse chemical species in the food. In general, researchers have discovered that increasing the electric field intensity and number of pulses causes bacteria to become inactive. Microorganisms may be killed as a result of pores forming in cell membranes. Pore formation mechanisms are not completely understood. Membrane rupture is triggered by an induced transmembrane potential around 1 Volt greater than the natural potential of the cell membrane, according to the dielectric rupture theory. The reversible or irreversible rupture of the cell membrane, referred to as electroporation, is influenced by factors such as the intensity of the electric field, the number of pulses, and the duration of the pulses.

Ultraviolet light has long been recognized as the most important element in bactericidal activity. Because of the poor penetration depth of the waves, it is mostly employed to sterilize air and thin liquid layers. When used in high doses, there is a strong propensity for flavor and odor to deteriorate before the items are sufficiently sterilized. Low-level radiation at precisely administered levels, on the other hand, can often enhance the shelf life of goods without compromising their quality. It is commonly known that UV light may be used to kill germs in water. This type of radiation is safer, more environmentally friendly, and less expensive to install and run than traditional chlorination. Unlike chlorine, it has no effect on the flavor of the water. High-intensity UV-C lamps have recently become accessible, with the ability to kill surface microorganisms on food.

Commercially, ultraviolet irradiation is used in the bactericidal UV lamps in a variety of food applications, including tenderizing meat, curing, and wrapping cheeses, preventing surface mold growth on baked items, and air purification in bottling and food processing facilities, as well as over pickle vats. Prerequisite programs are the foundation of any HACCP system; without them, food safety cannot be guaranteed. These systems establish methods for dealing with operational issues and ensuring public health. GMPs, sanitation standard operating procedures (SSOPs), water safety control, receiving, storage, and shipping control, pest and waste management, supplier control, trace and recall programs, equipment calibration, and personnel training are all examples of these procedures. The implementation of these precursor programs, which are often not included in a HACCP system and are rarely referred to as essential control points, is crucial to the development of a successful HACCP system (CCPs).

The dairy plant should be situated in such a way that it has a consistent and stable supply of potable water. It should also make garbage processing and disposal from the location as simple as possible. Furthermore, the climate of the place should be considered: strong winds or high temperatures would surely impair the product's quality. The facility's surroundings should be clean and pleasant for the employees. In general, the surrounding area should be neat, with no unnecessary collection of goods such as boxes, equipment, or machinery where pests could easily hide (FAO, 1990).

The plant's interior should be constructed to prevent contamination and ensure food product safety. Cleaning and sanitation operations should be made easier. The floors of dairy processing factories should be built of acid-resistant, nonslippery material. Tiles should be used to cover the walls. It is not a good idea to have equipment in contact with the walls or the floor. Stainless steel with no sharp edges should be used in engineering and manufacturing. It should be built and placed in such a way that cleaning and sanitation are simple and effective. The proper ventilation of the manufacturing plants ensures a healthy working environment for the employees. It must also be constructed in such a way as to prevent contamination. Positive pressure should be employed in processing regions to prevent airflow from raw material areas to finished goods. Additionally, processing, receiving, and storage rooms should have adequate lighting.

There must be restrooms for the employees, as well as suitable and accessible hand-washing stations. To avoid cross-contamination, water taps should be operated electrically. Manufacturing industry waste pipes should be separated from human sewers. The processing plant floors should be sloped to allow for waste drainage and effective plant cleaning. Finally, preventive maintenance of equipment, particularly that utilized for milk thermal treatment, should be carried out. Production facilities should be designed such that the entire process can be closely monitored. The dairy sector is concerned about cleanliness and sanitation since milk can easily be contaminated. Cleaning effectively can only be accomplished through a well-designed program that is performed as part of day-to-day operations. The employees in charge of implementing and overseeing the cleaning and sanitation operations should be educated. Furthermore, the program's efficacy and efficiency should be reviewed on a regular basis.

Finally, make a record of the cleaning methods. Food should be handled in a way that ensures its safety not only at receiving, but also during raw material storage and end-product distribution. The quality and hygiene of raw materials should be examined upon receipt, and any accompanying papers should be recorded and filed. The date, supplier, temperature, product condition, and certificates of analysis should all be recorded for arriving material. Raw materials and finished goods should be stored in a completely cleaned, pest-free, and tidy location with controlled environmental variables such as temperature and humidity.

One of the essentials of a food manufacturing process should be a pest control program. Documentation of the pest control operator's capability and liability, as well as the techniques and chemicals used, as well as the location of the pesticides, is required. A waste management program for the dairy industry is required. Waste management will eliminate pollution of the environment and water caused by garbage or hazardous chemical wastes. The chemical control program guarantees that harmful chemical substances are removed from food. Sanitizers, insecticides, mycotoxins, allergens, refrigerants, solvents, and acids are examples of chemical risks present in the dairy sector. These materials may interact with the raw materials, components, additives, and equipment, resulting in unfavorable outcomes. Food processing, cleaning, sanitation, and pest control all need the use of chemicals. Chemical control processes are often concerned with the receiving, storage, use, and disposal of chemicals. The supplier control program gives the organization a tool to evaluate raw material, ingredient, and service sources. The HACCP system might easily fail to give safety assurance without an efficient supplier control program. The key focus is the description of the industry's requirements for suppliers. These include the supplier's HACCP program, certificates of analysis, and documents of precursor programs that have been implemented (Manas et al., 2005).

From the receipt of raw materials and ingredients to the distribution of finished goods, the trace and recall program is utilized to keep track of food products. This program allows the corporation to recall products that are defective or harmful to human health. The construction of a successful withdrawal program comprises documented recall processes that keep detailed records of the process and the people responsible. This would contain data on raw ingredients, finished product code, consumer complaints, and how recalled products should be handled. All food processing equipment should be standardized and calibrated. Equipment that influences food safety, such as thermometers and pH meters, must be calibrated on a regular basis. Certificates issued by national or international organizations serve as assurances of trustworthiness. In many regions of the world, genetically engineered organisms have already found their way into the food supply. Various microorganisms, primarily bacteria and fungus, have been genetically manipulated to produce more proteins, amino acids, and commercial compounds. The finding of naturally existing or mutagenesis-induced mutant microbial strains was the mainstay of early research in this field. These mutant genotypes were frequently inhibited in specific metabolic pathways or exhibited increased quantities of a crucial rate-limiting enzyme, resulting in their metabolic output being directed toward the targeted outcome. For academics, such mutants were useful biological tools, and for the fermentation industry, they were a lucrative commercial asset (Maisanaba et al., 2016).

From the receipt of raw materials and ingredients to the distribution of finished goods, the trace and recall program is utilized to keep track of food products. This program allows the corporation to recall products that are defective or harmful to human health. The construction of a successful withdrawal program comprises documented recall processes that keep detailed records of the process and the people responsible. This would contain data on raw ingredients, finished product code, consumer complaints, and how recalled products should be handled. All food processing equipment should be standardized and calibrated. Equipment that influences food safety, such as thermometers and pH meters, must be calibrated on a regular basis. Certificates issued by national or international organizations serve as assurances of trustworthiness. In many regions of the world, genetically engineered organisms have already found their way into the food supply. Various microorganisms, primarily bacteria and fungus, have been genetically manipulated to produce more proteins, amino acids, and commercial compounds. The finding of naturally existing or mutagenesis-induced mutant microbial strains was the mainstay of early research in this field. These mutant genotypes were frequently inhibited in specific metabolic pathways or exhibited increased quantities of a crucial rate-limiting enzyme, resulting in their metabolic output being directed toward the targeted outcome. For academics, such mutants were useful biological tools, and for the fermentation industry, they were a lucrative commercial asset (Heldman, 2013).

One of the most important new food processing technologies in recent years is cold plasma. Plasma is made by partially or completely ionizing a gas. Cold plasma discharges used in food processing operate at temperatures close to room temperature, unlike hot plasmas like candle flames and welding arcs. Plasma systems with a high enough density of reactive chemicals can sterilize food and/or food contact surfaces without heating or otherwise harming the treated product. Cold plasma has been found to be a generally effective, flexible surface-sanitizing method against pathogenic bacteria, viruses, and parasites. Cold plasma inactivates foodborne pathogens by producing reactive chemical species and/or UV light to damage cell membranes, DNA, and other cellular components. Cold plasma is appealing for clean label and organic applications since these antibacterial modes of action are derived from ionized gases rather than the application of external chemical sanitizers. Cold plasma has appeal as a low-carbon footprint, chlorine-free, and sustainable method because it uses only air and energy as inputs. Processors of fresh and fresh-cut vegetables are particularly interested in direct cold plasma treatment because it is waterless and nonthermal.

Chlorine dioxide (ClO_2) is becoming a more commonly used sanitizer. ClO_2 can reduce bacterial infections on foods and food contact surfaces by more than 99.99% when used as an aqueous solution. ClO_2 can inactivate pathogens in hard-to-reach areas on fruits and vegetables when used in gaseous form. The interaction of sodium chlorite salt with an organic acid produces ClO2 gas.

Pulsed electric field (PEF) technology has been used to destroy suspended pathogenic bacteria in liquid foods such as juices and liquid eggs as a food processing method. The product is conveyed between PEF electrodes in a liquid suspension conveyor system to treat solid foods. Pretreating roots and tubers like white potatoes, sweet potatoes, etc., for future processing into fries, chips, and other starchy products is one application. PEF electroporated the vegetable cells, releasing internal chemicals and decreasing sugars. This change minimizes the degree and likelihood of overbrowning during frying. Sugar and minerals can be extracted more efficiently from sugar beet fibers, peels, and stems using the same type of sugar-release event. PEF has been tested for better shelf life and preservation of semiliquid food products, as well as better extraction of bioactive and nutraceutical cellular components like antioxidants, essential oils, and algal protein, among other things.

Advanced food processing technologies, such as those detailed in this article, will improve the safety and quality of a wide range of foods and prepared food products, as well as enable for new product development and process optimization applications. Many of these technologies have a specific set of commodities for which they are most efficient or make the most financial sense to use. Foods with a specific moisture content, shelf life, or composition might be appropriate product requirements. When employed in synergistic protocols with each other and mild heat, these modern technologies broaden the range of pathogens that can be efficiently addressed. Consumer education and marketing will be critical factors in determining whether new and unfamiliar technologies are accepted or not. Advanced food processing technology can preserve customers' health while also providing new opportunities for growers and processors with effective applications for major product lines to address high-impact food safety concerns.

The previous 25 years of research and development devoted to milk and milk products have resulted in significant improvements in processing plant mechanization, automation, and sanitation, as well as quality and safety, shelf-life extensions, and new product releases that have increased customer choice and convenience. The consumer has benefited considerably in ways that are not often obvious, such as the improved quality and shelf life of milk and milk products. Consolidation has undoubtedly occurred in the milk and milk products industries over the last quarter-century, as evidenced by the presenting of production statistics. Because of consolidation, developments in mechanization and automation have been made to the point that today's milk processing plant would be unrecognizable to practitioners 25 years ago.

Vanilla, strawberry, chocolate, banana, and coffee-flavored milks have been increasingly popular in recent years. Children-friendly flavors have also been introduced. Many processors also provide packaged, single-service size milkshakes, which comprise milk, increased milk solids, sugar, and stabilizers for thickening and aeration when shaken, to compete at the convenience store with other beverage options. These shakes feature flavors that are similar to flavored milks. Limited quantities of milk-based beverages, which blend milk with other beverages such as fruit juice, are available. Although tempting, stabilizing milk protein, particularly casein micelles, outside of its typical pH and ionic strength range poses a significant technological barrier.

In 2003, 5,333 million liters of standard, light, and low-fat ice cream were produced in the United States. The frozen dairy desserts sector uses 10% of total milk production and 16% of made milk output in the United States. Production has increased from 4,591 million liters in 1980 to 3,104 million liters in 1955. In 2003, Canada produced 380 million liters of frozen dairy desserts, with 79% being ice cream and light ice cream, 9% being ice milk, and 4% being frozen yogurt. The number of facilities making hard ice cream in the United States demonstrates the trend of consolidation. In 2003, 400 manufacturers were registered, compared to 950 in 1980 and 3,500 in 1955. Canada and Europe have seen similar patterns. Consolidation, like fluid milk, has enabled advancements in mechanization and automation, as well as plant sanitation and cleanliness.

Bibliography

1. Afrc, R. F. (1989). Probiotics in man and animals. Journal of Applied Bacteriology, 66(5), 365–378. doi:10.1111/j.1365–2672.1989.tb05105.x.
2. Alvarez-Suarez, J. M., Mazzoni, L., Forbes-Hernandez, T. Y., Gasparrini, M., Sabbadini, S., & Giampieri, F. (2014). The effects of preharvest and postharvest factors on the nutritional quality of strawberry fruits: A review. Journal of Berry Research, 4(1), 1–10. doi:10.3233/jbr-140068.
3. Appendini, P., & Hotchkiss, J. H. (2002). Review of antimicrobial food packaging. Innovative Food Science & Emerging Technologies, 3(2), 113–126. doi:10.1016/s1466–8564(02)00012-7.
4. Arakawa, K., Kawai, Y., Ito, Y., Nakamura, K., Chujo, T., Nishimura, J., & Saito, T. (2010). HPLC purification and reevaluation of chemical identity of two circular bacteriocins, gassericin A and reutericin 6. Letters in Applied Microbiology, 50(4), 406–411. doi:10.1111/j.1472–765x.2010.02810.x.
5. Ávila, M., Gómez-Torres, N., Delgado, D., Gaya, P., & Garde, S. (2017). Industrial-scale application of Lactobacillus reuteri coupled with glycerol as a biopreservation system for inhibiting Clostridium tyrobutyricum in semihard ewe milk cheese. Food Microbiology, 66, 104–109. doi:10.1016/j.fm.2017.04.010.
6. Ayala-Zavala, J. F., Oms-Oliu, G., Odriozola-Serrano, I., González-Aguilar, G. A., Álvarez-Parrilla, E., & Martín-Belloso, O. (2007). Bio-preservation of fresh-cut tomatoes using natural antimicrobials. European Food Research and Technology, 226(5), 1047–1055. doi:10.1007/s00217-007-0630-z.
7. Bleecker, A. B., & Kende, H. (2000). Ethylene: A Gaseous Signal Molecule in Plants. Annual Review of Cell and Developmental Biology, 16(1), 1–18. doi:10.1146/annurev.cellbio.16.1.1.

8. Bunger, A., Moyano, P. C., Vega, R. E., Guerrero, P., & Osorio, F. (2004). Osmotic Dehydration and Freezing as Combined Processes on Apple Preservation. Food Science and Technology International, 10(3), 163–170. doi:10.1177/1082013204044828.
9. Burt, S., & Reinders, R. (2003). Antibacterial activity of selected plant essential oils against Escherichia coli O157:H7. Letters in Applied Microbiology, 36(3), 162–167. doi:10.1046/j.1472–765x.2003.01285.x.
10. Chen, H., Chiou, C., & Chang, S. (2017). Comparison of methylparaben, ethylparaben, and propylparaben adsorption onto magnetic nanoparticles with phenyl group. Powder Technology, 311, 426–431. doi:10.1016/j.powtec.2017.01.060.
11. Chibane, L. B., Degraeve, P., Ferhout, H., Bouajila, J., & Oulahal, N. (2018). Plant antimicrobial polyphenols as potential natural food preservatives. Journal of the Science of Food and Agriculture, 99(4), 1457–1474. doi:10.1002/jsfa.9357.
12. Chon, S., & Kim, Y. (2004). Herbicidal Potential and Quantification of Suspected Allelochemicals from Four Grass Crop Extracts. Journal of Agronomy and Crop Science, 190(2), 145–150. doi:10.1111/j.1439–037x.2004.00088.x.
13. Chu, F. S. (1991). Mycotoxins: Food contamination, mechanism, carcinogenic potential, and preventive measures. Mutation Research/Genetic Toxicology, 259(3–4), 291–306. doi:10.1016/0165–1218(91)90124-5.
14. Deegan, L. H., Cotter, P. D., Hill, C., & Ross, P. (2006). Bacteriocins: Biological tools for bio-preservation and shelf-life extension. International Dairy Journal, 16(9), 1058–1071. doi:10.1016/j.idairyj.2005.10.026.
15. Delaquis, P. (2002). Antimicrobial activity of individual and mixed fractions of dill, cilantro, coriander and eucalyptus essential oils. International Journal of Food Microbiology, 74(1–2), 101–109. doi:10.1016/s0168–1605(01)00734-6.
16. Delcarlo, S. B., Parada, R., Schelegueda, L. I., Vallejo, M., Marguet, E. R., & Campos, C. A. (2019). From the isolation of bacteriocinogenic LAB strains to the application for fish paste biopreservation. Lwt, 110, 239–246. doi:10.1016/j.lwt.2019.04.079.
17. Delves-Broughton, J., Blackburn, P., Evans, R. J., & Hugenholtz, J. (1996). Applications of the bacteriocin, nisin. Antonie Van Leeuwenhoek, 69(2), 193–202. doi:10.1007/bf00399424.
18. Demirdöven, A., & Baysal, T. (2008). The Use of Ultrasound and Combined Technologies in Food Preservation. Food Reviews International, 25(1), 1–11. doi:10.1080/87559120802306157.
19. Dušková, E., Dušek, K., Indrák, P., & Smékalová, K. (2016). Postharvest changes in essential oil content and quality of lavender flowers. Industrial Crops and Products, 79, 225–231. doi:10.1016/j.indcrop.2015.11.007.

20. Eklund, T. (1980). Inhibition of Growth and Uptake Processes in Bacteria by Some Chemical Food Preservatives. Journal of Applied Bacteriology, 48(3), 423–432. doi:10.1111/j.1365–2672.1980.tb01031.x.

21. Elotmani, F., & Assobhei, O. (2004). In vitro inhibition of microbial flora of fish by nisin and lactoperoxidase system. Letters in Applied Microbiology, 38(1), 60–65. doi:10.1046/j.1472–765x.2003.01441.x.

22. Embarek, P. K. (1994). Presence, detection and growth of Listeria monocytogenes in seafoods: A review. International Journal of Food Microbiology, 23(1), 17–34. doi:10.1016/0168-1605(94)90219-4.

23. Esimbekova, E. N., Asanova, A. A., Deeva, A. A., & Kratasyuk, V. A. (2017). Inhibition effect of food preservatives on endoproteinases. Food Chemistry, 235, 294–297. doi:10.1016/j.foodchem.2017.05.059.

24. Ezhilarasi, P. N., Karthik, P., Chhanwal, N., & Anandharamakrishnan, C. (2012). Nanoencapsulation Techniques for Food Bioactive Components: A Review. Food and Bioprocess Technology, 6(3), 628–647. doi:10.1007/s11947-012-0944-0.

25. Gálvez, A., Abriouel, H., López, R. L., & Omar, N. B. (2007). Bacteriocin-based strategies for food biopreservation. International Journal of Food Microbiology, 120(1–2), 51–70. doi:10.1016/j.ijfoodmicro.2007.06.001.

26. Gálvez, A., López, R. L., Pulido, R. P., & Burgos, M. J. (2014). Biopreservation of Egg Products. Food Biopreservation Springer Briefs in Food, Health, and Nutrition, 71–73. doi:10.1007/978–1-4939–2029–7_6.

27. García, P., Rodríguez, L., Rodríguez, A., & Martínez, B. (2010). Food biopreservation: Promising strategies using bacteriocins, bacteriophages and endolysins. Trends in Food Science & Technology, 21(8), 373–382. doi:10.1016/j.tifs.2010.04.010.

28. Gokoglu, N. (2018). Novel natural food preservatives and applications in seafood preservation: A review. Journal of the Science of Food and Agriculture, 99(5), 2068–2077. doi:10.1002/jsfa.9416.

29. Gómez-Sala, B., Herranz, C., Díaz-Freitas, B., Hernández, P. E., Sala, A., & Cintas, L. M. (2016). Strategies to increase the hygienic and economic value of fresh fish: Biopreservation using lactic acid bacteria of marine origin. International Journal of Food Microbiology, 223, 41–49. doi:10.1016/j.ijfoodmicro.2016.02.005.

30. Han, Y., Jia, X., Liu, X., Duan, T., & Chen, H. (2010). DLLME Combined with GC–MS for the Determination of Methylparaben, Ethylparaben, Propylparaben and Butylparaben in Beverage Samples. Chromatographia, 72(3–4), 351–355. doi:10.1365/s10337-010-1646-2.
31. Heldman, D. R. (2013). Food Preservation Process Design. Food Engineering Series Advances in Food Process Engineering Research and Applications, 489–497. doi:10.1007/978-1-4614-7906-2_24.
32. Hugas, M. (1998). Bacteriocinogenic lactic acid bacteria for the biopreservation of meat and meat products. Meat Science, 49. doi:10.1016/s0309–1740(98)90044-4.
33. Hultin, H. O. (1994). Oxidation of lipids in seafoods. Seafoods: Chemistry, Processing Technology and Quality, 49–74. doi:10.1007/978–1-4615–2181–5_5.
34. Huss, H. H., Jeppesen, V. F., Johansen, C., & Gram, L. (1995). Biopreservation of Fish Products—A Review of Recent Approaches and Results. Journal of Aquatic Food Product Technology, 4(2), 5–26. doi:10.1300/j030v04n02_02.
35. Imran, M., Revol-Junelles, A., Paris, C., Guedon, E., Linder, M., & Desobry, S. (2015). Liposomal nanodelivery systems using soy and marine lecithin to encapsulate food biopreservative nisin. LWT – Food Science and Technology, 62(1), 341–349. doi:10.1016/j.lwt.2014.12.046.
36. Isolauri, E., Salminen, S., & Ouwehand, A. C. (2004). Probiotics. Best Practice & Research Clinical Gastroenterology, 18(2), 299–313. doi:10.1016/j.bpg.2003.10.006.
37. Jhandai, P., Jadhav, V. J., & Gupta, R. (2019). Bio-preservation of Foods: A Review. European Journal of Nutrition & Food Safety, 164–174. doi:10.9734/ejnfs/2019/v11i430159.
38. Junior, G. B., Abreu, M. S., João Gabriel Dos Santos Da Rosa, Pinheiro, C. G., Heinzmann, B. M., Caron, B. O., & Barcellos, L. J. (2018). Lippia alba and Aloysia triphylla essential oils are anxiolytic without inducing aversiveness in fish. Aquaculture, 482, 49–56. doi:10.1016/j.aquaculture.2017.09.023.
39. Kang, H., Kim, M., Hwang, U., Jeong, C., & Lee, J. (2019). Effects of methylparaben, ethylparaben, and propylparaben on life parameters and sex ratio in the marine copepod Tigriopus japonicus. Chemosphere, 226, 388–394. doi:10.1016/j.chemosphere.2019.03.151.

40. Kim, T., Tran, T. H., Hwang, S. Y., Park, J., Oh, D. X., & Kim, B. (2019). Crab-on-a-Tree: All Biorenewable, Optical and Radio Frequency Transparent Barrier Nanocoating for Food Packaging. ACS Nano, 13(4), 3796–3805. doi:10.1021/acsnano.8b08522.

41. Laursen, B. G., Leisner, J. J., & Dalgaard, P. (2006). Carnobacterium Species: Effect of Metabolic Activity and Interaction with Brochothrix thermosphacta on Sensory Characteristics of Modified Atmosphere Packed Shrimp. Journal of Agricultural and Food Chemistry, 54(10), 3604–3611. doi:10.1021/jf053017f.

42. Leistner, L. (2000). Basic aspects of food preservation by hurdle technology. International Journal of Food Microbiology, 55(1–3), 181–186. doi:10.1016/s0168-1605(00)00161-6.

43. Lillard, H. S. (1994). Effect of Trisodium Phosphate on Salmonellae Attached to Chicken Skin. Journal of Food Protection, 57(6), 465–469. doi:10.4315/0362–028x-57.6.465.

44. Liu, S., & Tsao, M. (2009). Inhibition of spoilage yeasts in cheese by killer yeast Williopsis saturnus var. saturnus. International Journal of Food Microbiology, 131(2–3), 280–282. doi:10.1016/j.ijfoodmicro.2009.03.009.

45. Maisanaba, S., Llana-Ruiz-Cabello, M., Gutiérrez-Praena, D., Pichardo, S., Puerto, M., Prieto, A. I., & Cameán, A. M. (2016). New advances in active packaging incorporated with essential oils or their main components for food preservation. Food Reviews International, 33(5), 447–515. doi:10.1080/87559129.2016.1175010.

46. Mañas, P., & Pagán, R. (2005). Microbial inactivation by new technologies of food preservation. Journal of Applied Microbiology, 98(6), 1387–1399. doi:10.1111/j.1365–2672.2005.02561.x.

47. Manual on Simple Methods of Meat Preservation (n.d.). Retrieved May 21, 2021, from http://www.fao.org/3/x6932e/X6932E00.htm.

48. Martínez-Graciá, C., González-Bermúdez, C. A., Cabellero-Valcárcel, A. M., Santaella-Pascual, M., & Frontela-Saseta, C. (2015). Use of herbs and spices for food preservation: Advantages and limitations. Current Opinion in Food Science, 6, 38–43. doi:10.1016/j.cofs.2015.11.011.

49. Mohapatra, A. R., & Jeevaratnam, K. (2019). Inhibiting bacterial colonization on catheters: Antibacterial and antibiofilm activities of bacteriocins from Lactobacillus plantarum SJ33. Journal of Global Antimicrobial Resistance, 19, 85–92. doi:10.1016/j.jgar.2019.02.021.

50. Naderi, N., House, J. D., Pouliot, Y., & Doyen, A. (2017). Effects of High Hydrostatic Pressure Processing on Hen Egg Compounds and Egg Products. Comprehensive Reviews in Food Science and Food Safety, 16(4), 707–720. doi:10.1111/1541–4337.12273.
51. Napier, D., & Combrink, N. (2006). Aspects of Calcium Nutrition to Limit Plant Physiological Disorders. Acta Horticulturae, 702, 107–116. doi:10.17660/actahortic.2006.702.13.
52. Németh, C., Mráz, B., Friedrich, L., Suhajda, A., Janzsó, B., & Balla, C. (2011). Microbiological measurements for the development of a new preservation procedure for liquid egg. Czech Journal of Food Sciences, 29(6), 569–574. doi:10.17221/362/2010-cjfs.
53. Nychas, G. J. (1995). Natural antimicrobials from plants. New Methods of Food Preservation, 58–89. doi:10.1007/978-1-4615-2105-1_4.
54. Parke, D. V., & Lewis, D. F. (1992). Safety aspects of food preservatives. Food Additives and Contaminants, 9(5), 561–577. doi:10.1080/02652039209374110.
55. Payne, K. D., RicoMunoz, E., & Davidson, P. M. (1989). The Antimicrobial Activity of Phenolic Compounds Against Listeria monocytogenes and Their Effectiveness in a Model Milk System. Journal of Food Protection, 52(3), 151–153. doi:10.4315/0362-028x-52.3.151.
56. Petrak, T., Medic, H., Novakovic, P., & Botka-Petrak, K. (2000). Bacteriological Contamination Of Egg Products After Thermal Preservation Processes. Acta Alimentaria, 29(4), 315–322. doi:10.1556/aalim.29.2000.4.1.
57. Prusky, D. (2011). Reduction of the incidence of postharvest quality losses, and future prospects. Food Security, 3(4), 463–474. doi:10.1007/s12571-011-0147-y.
58. Salas, M. L., Mounier, J., Valence, F., Coton, M., Thierry, A., & Coton, E. (2017). Antifungal Microbial Agents for Food Biopreservation—A Review. Microorganisms, 5(3), 37. doi:10.3390/microorganisms5030037.
59. Schwenninger, S. M., Meile, L., & Lacroix, C. (2011). Antifungal lactic acid bacteria and propionibacteria for food biopreservation. Protective Cultures, Antimicrobial Metabolites and Bacteriophages for Food and Beverage Biopreservation, 27–62. doi:10.1533/9780857090522.1.27.

60. Sebranek, J. G. (2009). Basic Curing Ingredients. Ingredients in Meat Products, 1–23. doi:10.1007/978-0-387-71327-4_1.

61. Sidhu, M. S., Sørum, H., & Holck, A. (2002). Resistance to Quaternary Ammonium Compounds in Food-Related Bacteria. Microbial Drug Resistance, 8(4), 393–399. doi:10.1089/10766290260469679.

62. Silva, H. D., Cerqueira, M. Â., & Vicente, A. A. (2011). Nanoemulsions for Food Applications: Development and Characterization. Food and Bioprocess Technology, 5(3), 854–867. doi:10.1007/s11947-011-0683-7.

63. Silva, M., & Lidon, F. (2016). Food preservatives – An overview on applications and side effects. Emirates Journal of Food and Agriculture, 28(6), 366. doi:10.9755/ejfa.2016-04-351.

64. Singh, V. P. (2018). Recent approaches in food bio-preservation – a review. Open Veterinary Journal, 8(1), 104. doi:10.4314/ovj.v8i1.16.

65. Stiles, M. E. (1996). Biopreservation by lactic acid bacteria. Antonie Van Leeuwenhoek, 70(2–4), 331–345. doi:10.1007/bf00395940.

66. Strohl, W. R. (2014). Antimicrobials. Microbial Diversity and Bioprospecting, 336–355. doi:10.1128/9781555817770.ch31.

67. Tadesse, F. (1991). Post-Harvest Losses of Fruits and Vegetables in Horticultural State Farms. Acta Horticulturae, (270), 261–270. doi:10.17660/actahortic.1991.270.31.

68. Tiwari, U., & Cummins, E. (2013). Factors influencing levels of phytochemicals in selected fruit and vegetables during pre and postharvest food processing operations. Food Research International, 50(2), 497–506. doi:10.1016/j.foodres.2011.09.007.

69. Trinetta, V., Vaidya, N., Linton, R., & Morgan, M. (2010). Evaluation of Chlorine Dioxide Gas Residues on Selected Food Produce. Journal of Food Science, 76(1). doi:10.1111/j.1750–3841.2010.01911.x.

70. Williams, N. T. (2010). Probiotics. American Journal of Health-System Pharmacy, 67(6), 449–458. doi:10.2146/ajhp090168.

71. Wong, A. C. (1998). Biofilms in Food Processing Environments. Journal of Dairy Science, 81(10), 2765–2770. doi:10.3168/jds.s0022–0302(98)75834-5.

72. Yan, J., Yang, R., Yu, S., & Zhao, W. (2021). The strategy of biopreservation of meat product against MRSA using lytic domain of lysin from Staphylococcus aureus bacteriophage. Food Bioscience, 41, 100967. doi:10.1016/j.fbio.2021.100967.

73. Zengin, N., Yüzbaşıoğlu, D., Ünal, F., Yılmaz, S., & Aksoy, H. (2011). The evaluation of the genotoxicity of two food preservatives: Sodium benzoate and potassium benzoate. Food and Chemical Toxicology, 49(4), 763–769. doi:10.1016/j.fct.2010.11.040.

INDEX

C

D

E

F